# Brand Protection, Security Labeling and Packaging

Technologies and strategies for optimum product protection

# Other Labels & Labeling books:

**ENCYCLOPEDIA OF LABEL TECHNOLOGY**
Michael Fairley

**THE HISTORY OF LABELS**
Michael Fairley and Tony White

**DIGITAL LABEL AND PACKAGE PRINTING**
Michael Fairley

**ENVIRONMENTAL PERFORMANCE AND SUSTAINABLE LABELING**
Michael Fairley and Danielle Jerschefske

**CONVENTIONAL LABEL PRINTING PROCESSES**
John Morton and Robert Shimmin

**LABEL DESIGN AND ORIGINATION**
John Morton and Robert Shimmin

**LABEL DISPENSING AND APPLICATION TECHNOLOGY**
Michael Fairley

**CODES AND CODING TECHNOLOGY**
Michael Fairley

**LABEL EMBELLISHMENTS AND SPECIAL APPLICATIONS**
John Morton and Robert Shimmin

**BRAND PROTECTION, SECURITY LABELING AND PACKAGING**
Jeremy Plimmer

**DIE-CUTTING AND TOOLING**
Michael Fairley

**MANAGEMENT INFORMATION SYSTEMS AND WORKFLOW AUTOMATION**
Michael Fairley

**SHRINK SLEEVE TECHNOLOGY**
Michael Fairley and Séamus Lafferty

**LABEL MARKETS AND APPLICATIONS**
John Penhallow

For the latest list please visit: **www.labelsandlabeling.com**

# Brand Protection, Security Labeling and Packaging

## Technologies and strategies for optimum product protection

**Jeremy Plimmer** FIMMM., DipM., MCIM.
Secretary - Product & Image Security Foundation

**Brand Protection, Security Labeling and Packaging**
Technologies and strategies for optimum product protection

First edition published 2016 by:
Tarsus Exhibitions & Publishing Ltd

Printed by CreateSpace, an Amazon.com company.

ISBN 978-1-910507-11-7

# Contents

While every care has been taken to ensure the information, charts, diagrams and illustrations in this publication are correct at the time of publishing it is possible that technology, specifications, markets and applications, or terminology may change at any time, or that the editor's or contributor's research or interpretation may not be regarded as the latest accepted guidance in some parts of the world of labels.

The publishers therefore cannot accept responsibility for any errors of interpretation or for any actions, decisions or practices that readers may take based on the publication content and would advise that the latest industry supplier specifications, standards, legislative requirements, performance guidelines, practices and methodology should always be sought before any investment or implementation is made.

# Preface

Printed labels and packaging are at the forefront of everyday international and national consumer trade and are also extensively used in industrial manufacturing and day-to-day commerce. Quite simply, they are an essential medium that is used to promote and market brands, to provide essential consumer health and safety information on products, to provide guidance on usage or dosage of medicines, to enable the assembly, traceability or tracking of goods and, increasingly, to provide links to websites or social networks.

In short, labels and packaging of all kinds are seen by millions of people worldwide each and every day. The information printed on them is widely trusted and is accepted as being accurate, safe and sure – even seen as a guarantee of brand or manufacturer provenance.

Sadly, organized crime gangs and criminals also know this, using their own counterfeit labels and packaging to pass off fake goods as genuine, to create tax or VAT fraud, to re-label inferior goods to appear as a higher quality, or to simply divert products from one part of the world to another. Indeed the ingenuity of the criminal mind has proved to be almost endless, often leading to products on the market that have health or safety concerns and, in some cases, leading to serious injury or even loss of life.

Fortunately, the label and packaging industries are also ingenious and work hard to create methods of combating counterfeiting, fraud, diversion, parallel trading, tampering, and much more using all kinds of security substrates, inks and converting processes, often combined with RFID, taggants, electronic article surveillance (EAS) techniques, unique codes, or holograms. There are literally hundreds of brand protection and security labeling solutions available today.

This book in the Label Academy series assesses the problems of brand protection and security labeling, discusses the many and varied solutions that can be used, describes when and how they can be incorporated – either on their own or in combination – and provides guidance on developing brand protection and security labeling systems and procedures. It should provide an invaluable resource and training aid for everyone that is either involved in producing or using all kinds of labels or packaging.

**Jeremy Plimmer M.C.I.M, Dip.M., F.I.M.M.M**
Secretary - Product & Image Security Foundation

# About the Label Academy

This book is part of the recommended study material for the Label Academy, a global training and certification program for the label industry. The Label Academy was created by the team behind Labels & Labeling magazine and the Labelexpo series of events.

The Academy consists of a series of self-study modules, combining free access to relevant articles and videos with paid text books (both printed and electronic). Once a student has completed a module, there is an opportunity to take an online test and earn a certificate.

It is expected that a Label Academy qualification will become a standard in the industry – for printers/converters, suppliers, brand owners and designers – and assist in providing a benchmark. In addition to its own training, the Label Academy will aim to become a resource provider to the many existing educational programs in the industry. Accredited training courses will be promoted through the Label Academy website and books will be provided at discounted rates.

The Label Academy concept was pioneered by industry expert Mike Fairley. This was in response to a reduction in the number of dedicated printing colleges and the need to standardize training across the world. The label industry also has its own specific training needs – it has some of the widest range of materials, printing processes and finishing solutions of any printing sector.

We are also working with other training experts and authors to ensure that the Label Academy provides up-to-date and relevant training material for the industry.

The Label Academy is supported by the key trade associations, including FINAT, TLMI and the LMAI.

**www.label-academy.com**

---

# Label Academy sponsors

Thank you to our founding sponsors, without whom this ambitious project would not have been possible:

### Cerm

Cerm designs business automation software solutions to meet the specific demands of flexo and digital narrow web printers. Using the latest technology, our team's focus is on innovation and continuous improvement.

Our automation solutions support each step in the printer's integrated workflow – from estimating to production, shipment and data collection – and provide the feature and functionality printers need to gain efficiency and improve profitability.

Cerm inspires collaboration and helps printers remain competitive in the market and deliver the best products possible. We are proud to sponsor the Label Academy and contribute to the future of the narrow web printing industry.

**www.cerm.net**

### Flint Group Narrow Web

Flint Group Narrow Web has the products, the solutions, and the technical experts to handle any print situation. Providing solutions for food packaging, sustainability, increased bottom line, efficiency, and uptime – delivering the basics needed to run a successful operation, and the expertise to go above and beyond to another level of success.

Our experts provide solutions to your printing problems with the innovative products and services that have made us an industry leader around the world. Wherever you are, we are – available to help you reach your business goals today and into the future.

Continuous improvement is paramount to Flint Group; we are proud to sponsor the Label Academy and the benefits it will bring to the future of our industry.

**www.flintgrp.com**

### Gallus Group

The Gallus Group with its production sites in Switzerland and Germany is a leader in the development, production and sale of narrow-web, reel-fed presses designed for label manufacturers. The machine portfolio is augmented by a broad range of screen printing plates (Gallus Screeny), globally decentralized service operations, and a broad offering of printing accessories and replacement parts. The comprehensive portfolio also includes consulting services provided by label experts in all relevant printing and process engineering tasks. The Gallus Group is a member of the Heidelberg Group and employs around 430 people, of whom 253 are based in Switzerland. The group headquarters is in St.Gallen, Switzerland.
**www.gallus-group.com**

### MPS Systems B.V.

Producing high-quality label printing depends on several factors; one of them is the operator of the press.

As a press machine builder since 1996, MPS Systems B.V. knows how important training and education on subjects like pre-press, label printing and finishing is. For label printers, it is critical that their operators keep up with pre-press and press developments in addition to label trends. Therefore, MPS sponsors the Label Academy, to advance operator's passion for printing, share expertise and help multiply benefits.

The MPS slogans of 'Printers First' and 'Technology with Respect' have always underlined the core philosophy of MPS from press design to operator satisfaction. We develop our presses with a strong focus on user-friendliness and respect for the press operator: Printers First.
**www.mps4u.com**

### HP Indigo

HP Indigo is a global leader in digital printing, with a broad portfolio of digital presses and workflow solutions. Indigo's proprietary Liquid Electrophotography (LEP) technology delivers exceptional print quality for the widest variety of applications including labels, flexible packaging, shrink sleeves and folding cartons. HP Indigo's digital presses match gravure print quality satisfying the most demanding brands.

A division of HP Inc.'s Graphics Solutions Business, Indigo serves customers in more than 122 countries, including many of the top label and packaging converters worldwide.
**www.hp.com/go/labelsandpackaging**

### UPM Raflatac

In a little more than three decades, UPM Raflatac has become one of the world's leading manufacturers of pressure sensitive label materials, developing and leveraging the latest innovations in adhesive technology. Our film and paper label stocks are used for product and information labeling across a wide range of end-uses – from pharmaceuticals and security to food and beverage applications.

We are an engineering driven company with industry-leading products known for their consistent high quality and top performance. We are also known for the high performing supply chain and undisputed leadership in the area of sustainability. UPM Raflatac's dedication to innovation, sustainability and top quality is matched only by our commitment to service excellence. We call it the Raflatouch.
**www.upmraflatac.com**

# About the author

**Jeremy Plimmer M.C.I.M, Dip.M., F.I.M.M.M**
**Secretary - Product & Image Security Foundation**
Jeremy Plimmer has 35 years of experience in the security related print and packaging industry working originally as sales development director for manufacturers of passports, high security print and labels and more recently as a consultant. He is a graduate member of the Chartered Institute of Marketing, a Fellow of the Institute of Minerals, Materials and Mining, chairman of the West Midlands Packaging Society as well as vice chairman of the East Midlands Packaging Society.

As an author and co-author of a number of specialist market research studies, some commissioned by single clients and others published on a multi-client basis, he has identified market opportunities, benchmarked products, analysed markets and predicted future growth opportunities in print related authentication & security products for clients in the US, Canada, France, Germany, Holland, Belgium, Israel, Sweden, South Africa and the UK. He also has experience working with the European Commission, the OECD and a number of universities as an expert evaluator of suppliers of new authentication technologies requesting research and development grants.

Jeremy also acts as secretary general of the Product & Image Security Foundation, an international forum for manufacturers, suppliers and users of security labels, tags, tickets, materials, systems and document and product/image security technologies. The Foundation acts as an introductory source for authentication suppliers and advises its members in the security print and packaging industries in the areas of brand protection, paper security, print security, counterfeit deterrence and intelligent packaging. It achieves these aims through publishing a bi-monthly newsletter and through regular speaking engagements at conferences and seminars.

# Chapter 1

# Introduction to security and product protection

All of us come into contact with security labels in one form or another during our daily lives. We may not notice them, but be assured that security labels and packaging play an important part in protecting brand owners and us, their customer, from any number of disturbing risks.

The major role of packaging is to protect, inform and contain a product during its life from manufacture through to point of use. This function is applicable for the vast majority of merchandise whether consumer or industrial product, no matter how or where it will be used.

Labeling and packaging play a major part in supplying information and instructions in order to convey directives to the user that identify the brand, provide useful data regarding origin and age, and often an indication that products have not been tampered with during the distribution cycle.

All goods can be loosely classified as 'assets', since products have a value and it is this value that can be at risk from threats such as counterfeiting, diversion, tampering and theft. In this first part of the

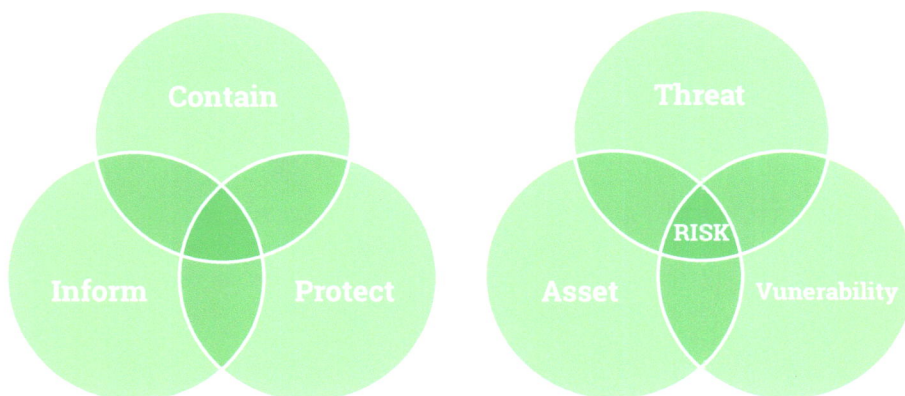

**Figure 1.1 -** The role of packaging and labeling is not just to 'contain', ' inform' & 'protect' but also to help manage certain risks within the supply chain

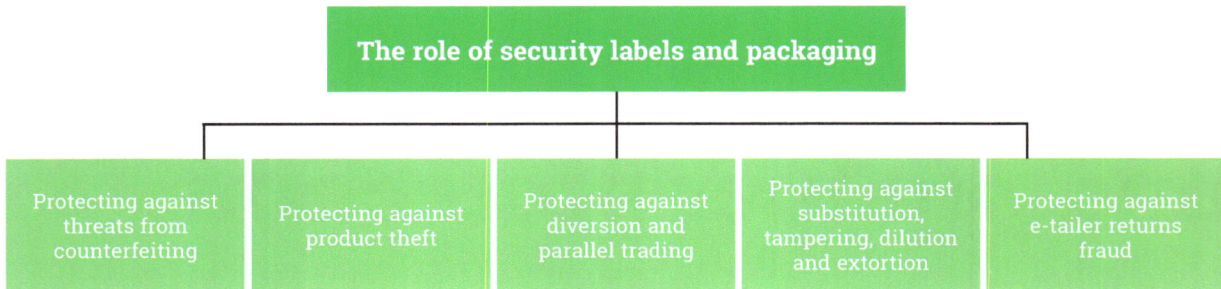

**Figure 1.2 -** Security labels and packaging can be used to protect 'assets'

study into security labels and packaging, the focus will be on how important it is to recognize where risks may be encountered in various product sectors, and how these can be mitigated through the use of a variety of techniques that are available to the informed labeling and packaging producer or supplier.

It is also important to understand that security labels and security packaging in isolation cannot protect products from attacks such as counterfeiting and tampering. What they can deliver though, is a cost effective indicator that a security threshold has been breached and that action is needed to address an important safety or operational issue that needs to be actioned by a specialized security team tasked with removing that threat.

All products have a value, some more than others. It is that value that comes under attack from criminals and lawbreakers who identify soft targets that can provide an easy source of income with little effort and little outlay.

## PROTECTING AGAINST THREATS FROM COUNTERFEITING

One of the biggest threats faced by manufacturers of industrial and consumer goods, including food and drink, is counterfeiting.

A surprising fact is that a product does not have to possess a high price tag to provide an attractive target to a counterfeiter. With luxury brands the value is perceived in exclusivity, materials and design. However, even mundane products such as toothpaste and household fragrance dispensers can attract unwanted

attention from fraudsters who know that these products contain inexpensive ingredients complemented by a household brand name. High product usage supports high prices through heavy advertising and an established pedigree. It is this added value that becomes an attractive target for the fraudster.

Counterfeiting is a crime that is global in scope and affects some twenty five market sectors across a wide range of goods. What is more, the market for fake products continues to grow and now touches an estimated 7 to 8 percent of trade or 1 trillion USD of revenue lost each and every year. (Some estimates put the figure as high as 1.8 trillion USD).

This type of crime is often considered 'victimless' and penalties can often seem insignificant when taken in context with the damage that can be done when fake medication, illicit liquor and bogus vehicle parts enter the supply chain.

Product related crime that encompasses counterfeiting and a number of other misdemeanors is growing by more than ten percent per annum, driven by the profits that can be made from cheap manufacturing. Economies in production are often supported by forced labor and distribution chains and methods that parallel those used to circulate narcotics and illegal weapons.

Indeed, product related crime often acts as a source of capital to fund even more heinous activities such as people trafficking, prostitution and terrorism, since it can generate large sums of cash quickly and with a minimum of risk.

Counterfeit products often pose an immediate

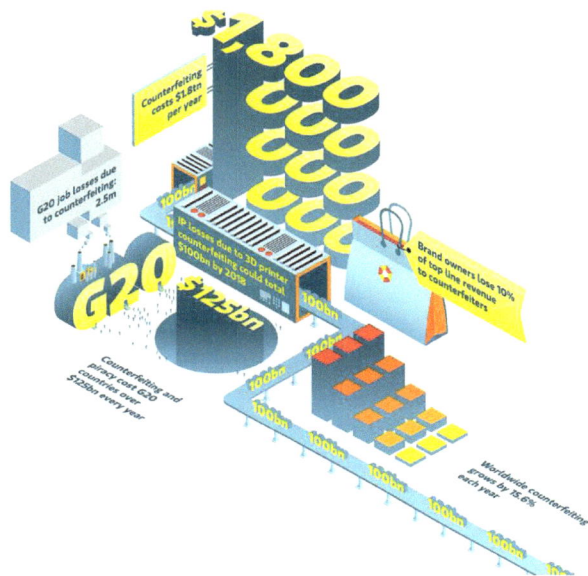

**Figure 1.3 -** The economic cost of counterfeiting to the global economy. *Source: NetNames*

danger to health since they are never manufactured to the same standards as authentic, proprietary merchandise. Most worrying, this trend is led by fake prescription drugs such as those used to treat cancer, heart disease and severe infections such as malaria and outbreaks of bird flu and suchlike.

Since all proprietary drugs are specifically researched and developed they carry with them a high cost that relates to the investment made in bringing them to market. These outlays require protection in the form of patents, which protect in law the right for the developer or brand owner to be shielded from reproductions of the same compound by competitive manufacturers.

When patents run their course, which is usually around twenty years, other manufacturers are then able to add these previously protected products to their portfolio.

Obviously, medication specially developed for both human and animal use becomes an attractive target to counterfeit. This is because the cost of these products is high and through substitution of

alternative non-performing ingredients criminals are able to place fake medication into the market at similar prices to authentic drugs and obtain immense profits from the operation.

This criminal activity has become so severe that governments have been forced to implement legislation that makes it mandatory for brand owners to apply markings to their authentic prescribed medical products so that they can be easily recognized as genuine by everyone in the supply chain.

This approach is by no means unique and a number of other industrial and consumer goods sectors of the market have, or are in the process of, developing systems that can be utilized to establish genuine product and provide an easily recognized indicator of authenticity for those tasked with policing the supply chain.

For instance there are many active consortiums of brand owners now operating in conjunction with specialized investigation teams supplied by established and recognized security operators who provide expert investigation staff who seek out suppliers and distributors of counterfeit products with a view to seizure and litigation.

It requires a high degree of skill and expertise to be able to track down and identify some fake products, since not all counterfeits are of poor quality or can readily be recognized as bogus-as they are on-sale from street vendors. Specialized teams of security investigators now operate in the tobacco, liquor and imaging supplies sectors. Likewise such squads are also available to supply service in the luxury goods, cosmetics, auto parts and electronic components sectors too.

Since packaging and labeling are common factors in most market sectors these items have become a de-facto carrier of indicia that can be useful in establishing the provenance of a product over and above the presence of a brand name, trade mark or registered design which can easily be copied.

By providing an easily recognized security printed, or applied indicia on product packaging and labeling, a brand owner is able to make available a standard authentication tool. This device enables investigation staff that are unskilled in meticulous product evaluation to make a judgement as to whether or not

● ● Pharmaceuticals, medications and veterinary products

● ● Clothing and footwear

● ● Cigarettes, tobacco and tobacco products

● ● Perfumes, colognes, toiletries and cosmetics

● ● Food and drink including alcoholic liquor

● Sporting goods and memorabilia

● Telecommunications equipment and accessories

● Computer equipment and supplies

● Imaging supplies, toners and ribbons

● Business and games software

● ● Dietary supplements and body building compounds

● ● Industrial and agricultural chemicals

● ● Automotive and aeronautical parts

● Purses, bags and travel accessories

● ● Headgear including safety helmets

● Sports clothing including official team strips

● Sound and video recordings

● Domestic entertainment equipment/consoles

● ● Toys and gadgets

● ● Electronic parts and components

**Figure 1.4 -** shows the major classifications of merchandise susceptible to counterfeit attack that can carry indicia on packaging and labels in order to prove provenance GREEN.
Where a risk to health and safety is present in counterfeit products of this type the RED indicator is used.

merchandise is indeed genuine.

Reference was made previously to health and safety, which are major issues when it comes to fake products being purchased and put into use unknowingly. Whilst purchasers may well be aware that they are buying a counterfeit product because it is self evident, many others may well be legitimately

duped into purchasing a fake, especially if this is acquired in a retail store, by direct mail or from, say, an internet pharmacy.

It makes good business sense for criminals to infiltrate their bogus products into the legitimate supply chain in order to mislead consumers into making an unsafe purchase. This can be accomplished through a number of devious tactics such as illegally employing legitimate staff to withdraw authentic product and replace this with imitations or by setting up 'official looking' websites or direct mail companies that mimic their real counterparts.

The major consumer 'at risk' sectors of this activity are alcoholic liquor, pharmaceuticals, automotive parts and cosmetics, as purchasing (and using) these fake products can prove to be dangerous as they often contain harmful ingredients or components. For instance bogus liquor, often found for sale on main street sole trader outlets can contain fatal amounts of methanol. Cosmetics, also found on sale in similar outlets can contain lethal doses of heavy or toxic metals. Fake prescription drugs do not contain pure or active ingredients and phony auto-parts such as brake pads can cause accidents because they fail to operate effectively.

Other less risky, but nevertheless potentially harmful contents, can be found in counterfeit cigarettes, clothing and footwear, electronic goods, dietary supplements, chemicals, headgear and children's toys.

Taking these sectors in turn it is not surprising to know that fake cigarettes and tobacco products often contain much more harmful ingredients than their real branded counterparts. Tobacco in fake products can be diluted with wood shavings, dried animal feces and other constituents that bear a similar texture and visual appearance to tobacco.

Clothing and footwear, safety headgear and toys are all products that require safety testing and compliance to standards that are designed to protect against accidental injury from fire, impact or ingestion in the case of children's toys, where lead paint can poison if a toy is chewed, or small parts can cause choking attacks.

Chemicals, both household and agricultural, are also popular targets for counterfeiters because they often carry high price tags and are highly concentrated or contain patented ingredients that have involved expensive research.

Finally, counterfeit electronic goods, in both finished usable and un-finished component form, can potentially cause harm to users as they are not manufactured to recognized safety standards such as UL (Underwriters Laboratory) in the U.S.A. and CE marking in the European Economic Area (EEA).

Safety marked products offer another attractive target for the counterfeiter as such fake items do not have to conform to rigorous safety standards.

Reference will be made later to a number of other commercial risks that may be addressed through the application of secure packaging.

## PROTECTING AGAINST THE THREAT OF PRODUCT THEFT

Product theft, from retail stores and also from within the supply chain, continues to cause substantial losses for businesses worldwide.

In some sectors of retail, theft from the display area and also from storerooms in the back of store, can amount to as much as two percentage points of lost turnover. Observant shoppers will notice the most at risk stores from the presence of security detection systems at each store entrance and exit. They take the form of tall gate-like structures placed each side of the gangway.

These detection systems are able to identify products that have not passed correctly through the checkouts, and therefore may be stolen. The system, known as Electronic Article Surveillance (EAS), operates through the presence of a label or tag embedded within the packaging of items at high risk from theft. Each label or tag has an embedded micro-circuit that is activated when products are placed into the retail supply chain or on display in store.

Each checkout, at stores equipped with these EAS theft detection systems, is equipped with a deactivation station. The EAS micro-circuit is only deactivated when paid for products are passed through the store checkout. Any products that contain an EAS tag that has not been deactivated will trigger an alarm as they pass through the security detection zone at the store exit. This alerts security

staffs who are then able to deal with the problem appropriately.

A number of EAS systems exist and all require specific security tags in label form to be present on at risk products in order to identify and deter theft. Some equipment used in this field depends on Radio Frequency Identification (RFID) and this technology

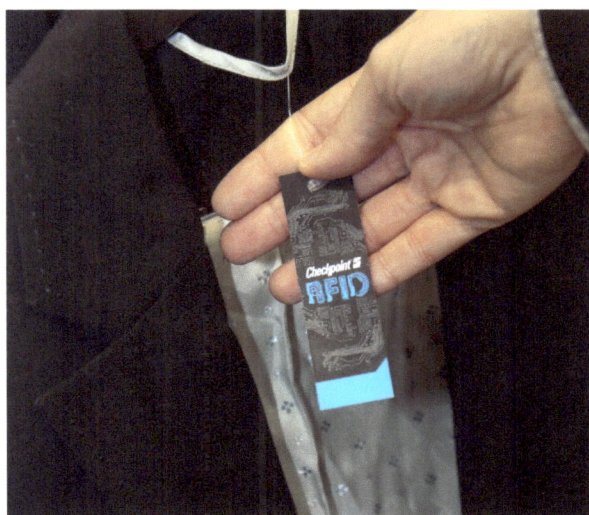

**Figure 1.5 -** An anti-theft RFID label or tag can be added to a product to identify stolen products in-store or in the supply chain. It may also be added to packaging either openly to deter theft, or furtively so as not to visually detract from established pack design

operates in a similar manner to contactless payments inasmuch as tags can be recognized and be activated or deactivated from a distance. Whilst contactless payments require a close proximity with the reading apparatus, EAS RFID can operate at distances enabling the technology to effectively cover the entrance of a store front, for instance.

It is possible to leverage a number of other benefits from RFID labels other than anti-theft devices, but more of this later.

The use of EAS security labels and tags is most commonly seen in self-service stores such as supermarkets, department stores, DIY centers and clothing and footwear outlets. The practice of adding EAS tags to liquor and luxury food products such as smoked salmon and expensive cuts of meat is also becoming prevalent as retailers attempt to detect and deter stock loss.

## DIVERSION AND PARALLEL TRADING THREATS NULLIFIED

So far reference has been made to how labels and packaging are useful platforms for authentication indicia and theft detection systems.

There are however, a number of other little known but nevertheless important risks, which may also be addressed through the presence of print related security systems or features that can be added to labels and packaging.

Some of these risks relate to the differences in selling prices for high profile brand-named products that exist from region to region throughout the world. The process is known as 'parallel trading' when regular goods are involved, and 'diversion' where excise duty is levied at differing rates.

These price differences may be because of taxation variances, or they may come about as brand owners seek to grasp the maximum value they can from a luxury good or indispensable product in a market that is opulent rather than deprived.

Take pharmaceutical products and medications as an example. These are often discounted in third world markets where the population is unable or unwilling to pay the same price that is demanded in the North American or wider European markets. The trick here is for wholesalers to buy in markets where products are less expensive (usually deprived or developing nations) and then move the goods to countries where prices are higher, thus making a tidy profit in the process.

Whilst this might be viewed as sharp practice by the legal brand owners, it is often difficult to deter and since a certain degree of value is lost to the manufacturer in the process, it requires some ingenuity in order to identify and halt the loss of profit and shareholder value that is eroded when parallel trading occurs.

The descriptive term 'parallel trading' relates to

products that flow into the market from official and unofficial sources in tandem. The official source is that route chosen by the brand owner and covers officially priced product. Traders and wholesalers then infiltrate the channel with cheaper authentic product sourced from elsewhere. There is also a risk of counterfeit product flowing within this chain too, so being able to identify legitimate goods within each discrete market is of major value to each brand owner.

A similar situation to this occurs when excisable products such as tobacco and liquor find their way into illegitimate distribution channels. In this case the brand owner is not the loser; rather it is national and regional governments that face the loss of taxation revenue when this illegal practice occurs.

Everybody will be aware that some products face taxation or excise duty charges as a revenue generating stream for either national governments or regional administrations. It's these taxes that help pay for services provided by the state. Rates of taxation are set centrally and often differ from country to country or region to region. In the USA, for instance, tobacco may be taxed at varying rates in adjoining states. In Europe liquor and wines are taxed at differing amounts, sometimes to deter unhealthy consumption, but also in order to generate as much revenue as possible for the administration in charge.

In economic terms products that are highly taxed and still find ready buyers are termed 'inelastic'. It's these products that face the danger of the highest levels of taxation, and also are at most risk of diversion from regimes that levy lower rates to administrations that tax much higher.

Originally, this law-breaking was termed smuggling, but more commonly today it is referred to as excise evasion. In order to overcome the threats relating to excise duty evasion and product diversion many regimes now mark excisable products with special security labels termed 'tax stamps' in order to show duty has been paid and also to help identify the source of origin of dutiable products.

Since both diversion and parallel trading result in significant loses for governments and brand owners there is a growing requirement to track and trace those products most at risk. Track and trace systems often take the form of numbered labels as well as

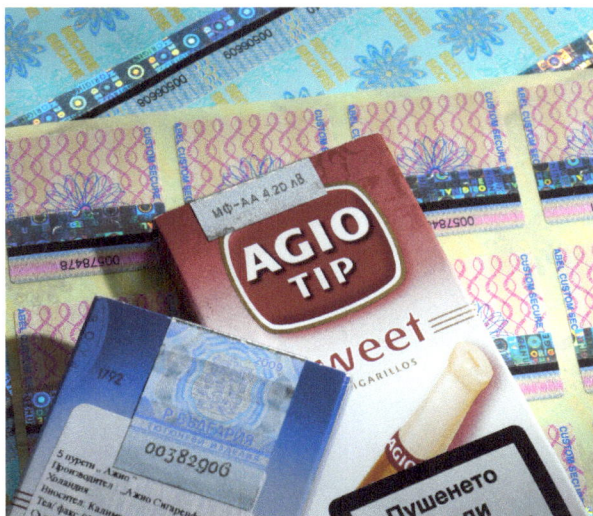

**Figure 1.6 -** Excise duty labels are useful tools in the fight against custom duty fraud and are affixed to at risk products either as self-adhesive stamps or as wet glue decals

special product marking codes that are useful in determining the authenticity and provenance of goods. Here then, packaging can be a useful platform for carrying identification markings that can be used to detect and deter illegal imports and unauthorized goods in retailers and distributors in the supply chain.

## PROTECTING AGAINST SUBSTITUTION, TAMPERING, DILUTION AND EXTORTION

Everyone recognizes the need for tamper evidence in all forms of packaging. Proof of first opening communicates with consumers the freshness of food and drink products and that a pack or jar remains safe against previous and maybe unsafe openings.

Tamper evident closures take many forms. Push button twist and open caps on jars, drop down tamper evident cap bands on bottles and weld seal zip locks on polyethylene bags. In the case of security related labels and printed packaging, focus is made on those closure seals that can be manufactured on label presses and other forms of traditional print equipment.

This is because printed tamper evident seals are

most useful when they can be made to satisfy a number of other major risks in addition to offering evidence of first opening, or that a product's packaging has remained safe and secure during transit and/or on a retailer's shelf.

A number of dangers exist in this 'sphere of influence' and these all relate to unlawful activities that encompass opening packs on supermarket shelves and part consuming product (grazing), then returning a re-closed pack back onto display for later consumption by an unaware purchaser. Other risks involve the refilling of discarded empty packs with

**Figure 1.7 -** Refilling is an ever present danger in the high value liquor market

inert or similar looking material through to 'spiking' product with harmful substances and then extorting a retailer or brand owner into paying for information on which products were affected and where, before they are consumed by an innocent purchaser.

Whilst this latter attack is thankfully very rare, it has been widely reported a number of times in the last decade or so and is treated in a similar manner to blackmail by the police.

Finally there is also the risk of dilution to guard

against. This occurs when packs are opened, the contents diluted with inert substances and the 'refilled' product is spread further (into counterfeit packs) making more profit for the supplier. This type of attack happens to liquids such as liquor, concentrated detergents and products supplied in bulk such as fertilizers, washing powders, chemicals, oils and paints.

As would be expected, a number of innovative solutions to these problems have been developed in

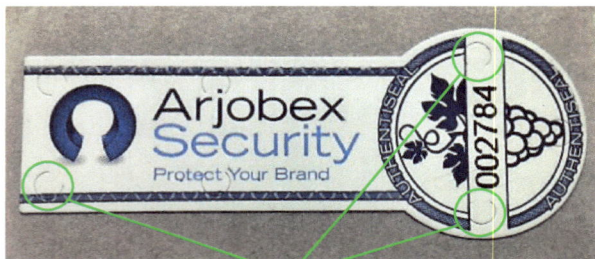

Tamper protecting perforations used on labels that deliver a 'seal' to screw caps for jars and bottles

**Figure 1.8 -** Various tamper evident constructions showing perforations and tamper indicating adhesives

order to address the risks associated with this type of product related crime.

The most simple construction is a self-adhesive 'bone label' with tamper strip perforations to deter and detect removal and re-affixing. This label is affixed (using wet glue applicators) over the pack closure and can only be removed through the destruction of its surface.

Making this construction more complex and

therefore more secure, involves adding tamper evident features to the adhesive by utilizing specialized substrates available from recognized material suppliers.

These 'printed' adhesives provide very good evidence of unauthorized opening.

More complicated versions of construction are available that allow for resealing after the pack has been opened. The integrity of the seal however, once opened, remains in a state of alert, with a clear warning message that is difficult to remove as it is formed from tinted, patterned adhesive.

An extremely cost effective approach for brand owners and designers is to add authentication devices to this method of securing a closure. If that process is adopted then two security threats may be accomplished simultaneously.

## RETURNS FRAUD – A GROWING PROBLEM FOR E-TAILERS

An e-tailer is a store or individual who does business only or primarily on the Internet. Both traditional main street retailers and their cousins on the world wide web suffer from returns fraud.

This type of fraud involves the purchase of expensive goods and then dismantling them and stripping out valuable components which are replaced with counterfeits or obsolete mechanisms that carry out the same function as the original.

For instance, high specification graphics cards on new computers can be replaced with outmoded parts. Re-clocked processors are also a target for this activity.

The main target for these types of attack focuses on industrial and military electronic components. This is a fast moving business and microelectronics are quickly outdated making their obsolete equivalents worth much less but so similar in visual appearance that they can be easily re-marked and sold for a higher amount.

Marking the authentic components and corresponding labeling and packaging with suitable security devices is an important part of providing clear identification features in this sector.

Likewise, there are similar risks within the e-commerce market place, especially where high cost

**Figure 1.9** - e-tailers are now a part of every consumer's environment

luxury or branded designer products are purchased with the intention to return them for refund after first use.

The trick here is that if you are planning to attend a wedding or similar event and want to impress everyone, you identify your chosen outfit on the net beforehand and then purchase in the usual way using your shopping cart and credit card. As long as your apparel and accessories are not marked you can return these later for a refund!

Alternatively, savvy (but dishonest) shoppers are now purchasing fake designer goods on internet auction sites and then buying their REAL counterparts from authentic suppliers who are then sent the counterfeit products as a 'return' for a refund.

Both these latter stings are serious matters for Internet-based retailers and also for mainstream retail that offers a (collect at store) or mail order delivery service.

In many cases the only sure fire method of identifying this type of fraud is by labeling products together with tamper proof hang tags that carry suitable security markings attached with closures that are also difficult to copy. Also these items should carry security features that enable quick identification to take place during returns inspection procedures.

This chapter contains a brief summary and selection of the most suitable applications for security enabled packaging and labeling. It is by no means exhaustive and during their day-to-day contact with customers, suppliers of these products will discover a number of other important opportunities for products of this type that can be manufactured on a printing press.

Counterfeiting is often viewed as a victimless crime. Nothing could be further from the truth. Criminals and terrorists do not 'partition' crime; to them it is homogeneous. Counterfeiting is often viewed as a relatively riskless activity for creating monies in order to finance more profitable endeavours such as people trafficking, drug distribution and attacks on democracy in order to destabilize governments.

These facts may be the reason why counterfeiting and its associated activities continue to be of interest to law enforcement agencies such as the FBI, Interpol and Europol. However, other players such as investigation agencies, intellectual property lawyers and trading standards organizations are also key performers in the reduction of activity in this field.

Since none of these groups are expert in identifying fakes across all sectors, security product marking, linked to effective control systems, is an important assistive tool when applied to labels and packaging.

# Chapter 2

---

# Brand protection – the role of security labeling and packaging

---

In the preceding chapter it was disclosed that branded products have a value which is often much higher than their worth in terms of materials, construction and distribution costs. When products contain very high value material content they present an attractive target to counterfeiters who can substitute high cost ingredients with much cheaper but similar looking alternatives.

---

In illustration of this point many pharmaceutical products contain constituents that are very expensive to make and involve a high degree of research and development and testing before they can be placed on the market. Introducing fakes of such products into the distribution stream allows criminals to substitute expensive valuable goods with inexpensive substitutes. In doing this they can almost double their investment since they can then resell the authentic goods for close to their real worth.

Conversely, even some relatively inexpensive products such as toothpaste, shoe polish and shampoos are attractive to the counterfeiter.

This is because with such items, which are eminently branded and recognized globally, their packaging holds more value than the contents, and since both contents and packaging are often easy to replicate they come under attack.

Hidden product value is therefore embedded in promotion costs in this instance, since the cost of building a successful global brand for toothpaste or shampoo relies on heavy advertising and investing in networks and distributors who also add cost to the product.

Of course counterfeiters do not have these costs to bear; they also have access to sources of cheap and often child labor so not only can they undercut the official holders of the brand, they can also make a tidy profit too.

Such undercutting practices, which are illegal in most countries that respect and oversee intellectual property (IP) law, require a systematic approach to enforcement and a coordinated response from a number of key players to ensure that counterfeit distribution is detected and discouraged.

In most developed countries trading standards are monitored by teams of lawyers and investigators who work under a number of banners. The Bureau of Consumer Protection in the USA and Trading Standards in the UK are illustrations of these organizations which have their counterparts elsewhere in Europe and other parts of the world.

The primary responsibility of these organizations is to ensure that individuals and businesses comply with national consumer protection laws and weights and measures.

These organizations have been established in order to provide consumers with a point of reference should they need to report an unfair or illegal trading practice such as counterfeiting, refilling or dilution frauds. They also periodically inspect and check goods in the consumer market in order to ensure they comply with legal trading guidelines and regulations.

In many cases brand owners work with local consumer protection teams in order to 'police' the market and investigate and prosecute anyone found selling fake products through any of the various routes to market. This covers street markets, flea markets, car boot sales in the UK, car trunk sales in the USA and sales through internet auction sites such as e-Bay and Alibaba.

Counterfeit goods also find their way into legitimate supply chains, so it is quite possible to purchase counterfeit products in local stores, on Main Street or even in supermarkets. Indeed there have been many recorded cases of counterfeit clothing, liquor, pharmaceuticals and cigarettes being sold through these 'official' channels through nationally and internationally recognized retail chains.

**Figure 2.1 -** Counterfeit headphones and purses openly on sale at a street market in Spain (2014)

Mostly the traders concerned are unaware that they are selling fakes since these have entered the legitimate supply chain without their knowledge and 'upstream' of the distribution network. It is also possible to find fake products within established, legal

industrial supply networks distributing chemicals, automotive parts and electronic components as well as an alarming number of other products. (see: Chapter 1 Figure 1.4)

The diagram (Figure 2.2) illustrates how the legitimate market for branded authentic products can be infiltrated by counterfeit goods.

Included in this schematic are a number of outcomes, that include satisfied customers who purchase authentic products through the bona fide channel (GREEN) and unwilling victims who unknowingly purchase counterfeit products thinking them to be genuine (ORANGE).

Of course there are many consumers who readily recognize they are purchasing counterfeit goods because they are involved in purchasing product at highly discounted prices on the street or from a flea market (BLACK).

Brand owners are of course keen to ensure that all distribution channels are monitored and as much counterfeit product is seized or removed from the supply chain before it reaches the market.

This simplified schematic also shows how legitimate gray market goods (also known as parallel traded products) find their way to market via unauthorized channels.

The flow chart (Figure 2.2) can be further complicated by the fact that criminals can attempt to disguise the infiltration of counterfeit goods into legitimate channels by mixing authentic and fake goods within the same batch.

Furthermore, the distribution channel for industrial products and components also provides an attractive target, and follows a similar pattern to that illustrated for consumer goods. Whilst companies may source materials and components directly from the manufacturer, many depend upon wholesalers and stockholders for their supplies of components and tooling.

These items are just as likely to be targets for copycat products as those within the domestic consumer channels.

At this point the complexity of the problem of fake products entering the supply chain for consumption by manufacturer and consumer alike will become unmistakable. The risks too are severe, especially for

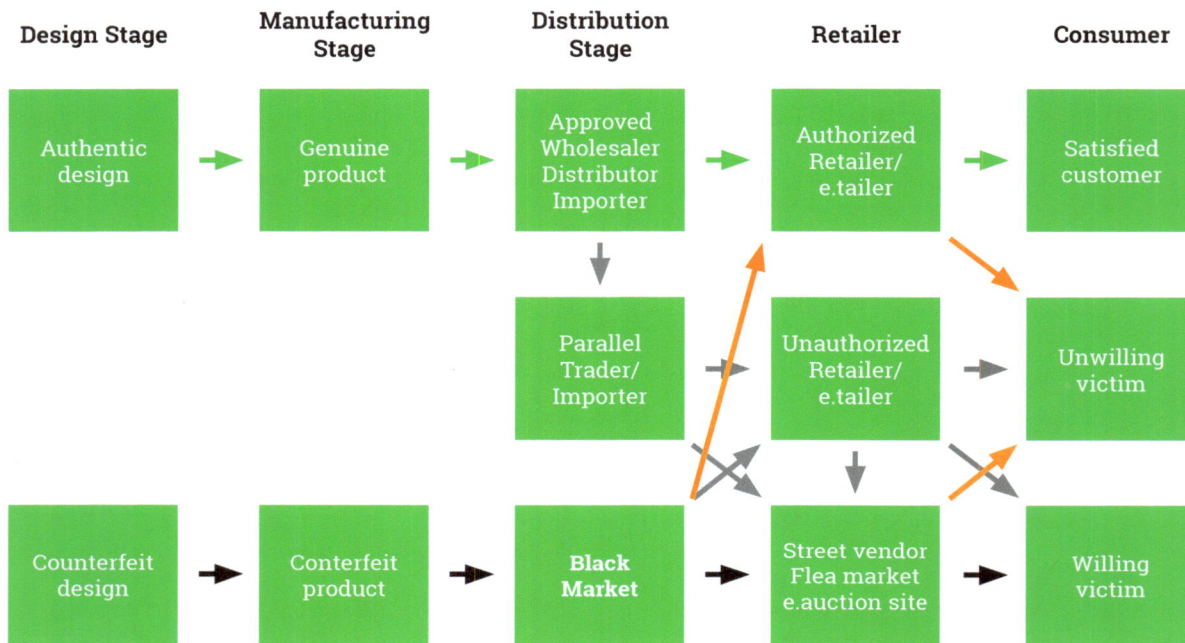

**Figure 2.2 -** The routes taken to market for authentic, parallel trade (gray) and counterfeit consumer products

goods that require safety certification or are required for medical purposes or human and animal consumption. Remember here that counterfeit veterinary products as well as pharmaceuticals are involved in this illegitimate trade.

The problem of reducing or eradicating counterfeit products from the supply chain is shared by a number of stakeholders.

Firstly brand owners have a critical interest in protecting their reputation as well as their profits. Governments, as was noted in the previous chapter, are keen to ensure that they do not lose out on revenue from taxable goods, and law abiding distributors must ensure that the products they sell are legitimate and they are safe from exposure to any IP infringement or warranty claims.

This last point returns our focus to intellectual property (IP) and its importance in establishing and protecting the rights of a manufacturing brand owner. (IP rights are common too in the service industry but

this sector is not affected by what can be termed 'hard product' fraud, neither is it a consumer of packaging).

## THE MARKING OF INTELLECTUAL PROPERTY IN ORDER TO DETER COPIES

Intellectual property covers a number of fields. Briefly this encompasses the power of the IP holder to enforce his right to use registered designs and processes in order to protect a brand against unfair competitive trading such as copy attacks and the theft of proprietary recipes/formulas and manufacturing methods and procedures[1].

**Trade Marks** are a recognized method of protecting a brand since they link a product to a known manufacturer and provide an established warranty of provenance and quality. Trade Marks can be likened to 'badges' inasmuch as they remain relatively constant over time in order for them to be readily recognized. Trade Marks may also involve

[1] This is a much simplified definition of IP and should not be taken out of context

discrete typography, logos, color or combinations of all of these attributes. They are recognized through the use of the TM symbol.

**Figure 2.3 -** The very first image to be registered as a UK Trade Mark was by Bass Brewery of Burton on Trent (1876)

**Registered Designs** cover the appearance of the whole or a part of a product resulting from the features of, in particular, the lines, contours, colors, shape, texture or materials of the product or its ornamentation. Registered design® also includes, in particular, packaging, get-up, graphic symbols, typographic type-faces and parts intended to be assembled in such a way that they display a uniqueness attributable only to that product, label or pack.

**Patents** provide the exclusive right of manufacture and sale of an invention for a prescribed number of years. These rights are delivered through patent offices worldwide and require a detailed analysis of the uniqueness of an invention or innovative step in process or manufacture before they are granted by a government and registered through its IP offices.

Before a patent is granted it requires detailed examination to ensure that it does not breach any previously published patents. This work is carried out by qualified patent attorneys. It is usual to mark all patented products with the word 'patented' and its applicable granted reference number.

**Copyright©** is a legal right of protection against copying and is applied to creative work in the form of writing, design and musical or visual performances. Most countries recognize copyright of any completed 'artistic' work without registration. Since copyright becomes relevant as soon as any artistic work is completed any copies made after this date will infringe the creator's right and therefore will become actionable. This publication is copyright and was protected as soon as it was completed.

It has been necessary to heavily abbreviate these definitions but they illustrate a number of important legal rights that are available to brand owners who find themselves exposed to counterfeit versions of their products in the market place.

Legal remedies exist in the form of injunctions; interim seizure orders, compensation, recall, disclosure and the publication of judicial decisions and all these are designed to ensure that infringers of IP may be called to account for their actions. Retribution is usually granted as financial damages and these can be costly for a defendant, such as a counterfeiter, if they are found to have breached any of these rights.

It should be recognized that legal cases that involve IP fraud can be very expensive to bring to court and that the burden of proof always resides with the prosecution. Often there is a very fine line drawn between the visual differences of a fake and an authentic product.

Any evidence given in a legal case will depend on proving a counterfeit is in fact what it is, a fake. It is only through examining differences between the authentic product and the fake that evidence can be shared in order to establish culpability.

In most cases this will require expert analysis and it is often more practical to rely on deviations to the packaging and labeling of counterfeits than say minute examination of the product itself. In illustration of this point imagine trying to prove that medication in the form of a tablet or capsule is counterfeit depending upon visual appearances such as shape, color or weight. In reality forensic evidence is needed and this can be impracticable and costly to acquire.

Similar problems exist with frauds involving liquids such as liquor and granulated products such as fertilizers.

These facts draw us to another important

conclusion when it comes to identifying fakes anywhere in the supply chain. Unless you are the brand owner or an expert in recognizing fakes within any particular category, how can you tell the difference between a fake and the real thing?

This is not just a difficult decision for the public at large, but also for customs officers and legal investigation teams working at pinch points in the supply chain in order to identify fake products.

Since brand owners often have limited investigative staff resources, they employ specialist investigation teams and the help of the World Customs Organization to examine consignments of goods on their behalf. Protecting the brand is an essential part of upholding brand value and reputation as has been revealed before. The high costs relating to IP supervision and protection are mostly carried by legal action and investigation. These can often be reduced by a little forethought and planning.

**Figure 2.4** - Packaging and labeling have an important role to play in identifying counterfeits

## SECURITY RELATED PRINT – THE ROLE OF LABELS AND PACKAGING IN PRODUCT PROTECTION

Since most products at risk from counterfeiting are packaged and labeled, these components offer a common point of reference across a whole variety of different merchandise. It is often more cost effective to teach those tasked with defending the brand to note – and easily identify – security features present in wrappers and decals, than it is to provide coaching in the intricacies of inspection for minute differences in product fabrication.

Therefore packaging and labeling have an important role to play in product protection since they offer a useful method of preventing unauthorized opening (tampering and refilling) as well as providing a bellwether of authenticity, since if expertly designed they can offer similar anti-counterfeit devices to those found on banknotes.

Security related print in the labels and packaging market has been largely driven over the last three decades by some of the design and security features found on banknotes.

Since banknotes have an intrinsically high value they are attractive to the counterfeiter. In order to deter and detect counterfeit attack the financial sector has developed a variety of security devices that can be useful in both prevention and discovery of fake currency.

At this point it would be useful to examine the strategy and tactics deployed in anti-counterfeit banknote design.

Initially, banknotes are required to meet two primary objectives. They should be easy to authenticate but difficult to reproduce or copy.

Taking the latter point first, any attempt to reproduce a banknote should be made as tough as possible for the counterfeiter who has a number of options open when launching a copy attack.

Firstly, a counterfeiter can attempt to photocopy a banknote and in order to prevent this happening, banknote designers have deployed special inks and intricate design mechanisms so that any copy is immediately recognizable as such and can be easily spotted.

Complex line
work forms
anti-scan pattern

Holographic
stripe (visual)

Color
changing ink

Raised tactile
printing sensitive
to touch

**Figure 2.5 -** *A few of the publically (overt) accessible security features found on a banknote (illustration courtesy of ECB)*

Secondly, a determined counterfeiter can try to scan or photograph a banknote using high resolution scanners or cameras. From that initial scan it is then possible to recreate a copy using digital printers in order to replicate the scan any number of times. The risks of scanning and photography attacks are avoided through the use of special inks and also anti-scan screens that either throw up a visible warning on any replication of the scan, or block out the use of the scan on a desktop printer or photocopier through currency anti-copy software that is inbuilt into the printer by the hardware provider.

Finally, it is always possible to attempt to recreate a banknote from scratch using origination software and traditional printing equipment. This type of attack is neutralized through the use of restricted materials and processes that make it next to impossible for a faker to obtain access to proprietary design software and security papers and inks that are used in the official production of currency by national banks and dedicated security printers.

Since currency counterfeiting is an ever present and persistent threat, it follows that any anti-counterfeit features that are deployed in order to reverse the trend be constantly updated as they become at risk of compromise. This provides a ready source of print related security devices that may be useful in brand protection applications since they are already established in daily currency verification and recognizable as such by the public at large and also those tasked with protecting the brand.

**RECOGNIZED PROCEDURES FOR IDENTIFYING COUNTERFEITS**

The recognized procedures for authenticating currency initially involve a number of human senses since handling banknotes is designed to be quick and uncomplicated.

A series of tests have evolved over time that provide a primary assessment of substantiation. These involve both sight and touch, mostly incorporating features such as watermarks, color changing inks, holograms and raised printing that can be sensed when running a finger over certain areas of the bill. Over time frequent users of paper currency such as shop assistants and bank tellers become accustomed to the 'feel' of authentic banknotes since they differ considerably from the paper that is available for use by a printer of counterfeit notes.

These initial tests are referred to as 'overt' or 'primary' practices where overt can be described as what can first be established and is obvious to a quick five or ten second check.

If an item of paper currency does not pass this first stage of certification and thereby becomes suspect, a second series of tests are available to establish legitimacy.

Such tests involve 'covert' or hidden features that reside with a banknote and can only be discovered by using some form of apparatus such as a loupe or other high magnification tool. The use of this device will reveal security techniques such as micro-printed lines of type and the use of fine line-work rather than screens (dots) for color rendition and shading. Covert tests can also include validation pens that are used to check banknote paper and inks that change color under specific frequencies of non-visible light such as those found in the ultra violet or infra-red part of the spectrum.

Finally, if none of the previous tests deliver conclusive results a third 'forensic' level of challenge is bought into play. Forensic testing as its title implies involves scientific procedures that can be used to establish the genuineness of a banknote beyond all doubt. Examinations will include the observation and presence (or not) of chemical markers, biological markers (DNA) and other proprietary features within the currency that can only be extracted through laboratory processes and examination using technology such as spectral analysis or microscopic scrutiny.

The labeling and packaging industry has now learned, through the advice and knowledge of the security related printing industry, to include difficult to copy security features taken from the fiscal market and include these as counterfeit deterrence mechanisms in their products.

During the past decade a wide variety of other innovative methods have been developed to detect and deter counterfeit attacks on branded goods. These will be covered in detail in the forthcoming chapters.

Preceding paragraphs have mentioned that security features embedded in print and packaging can be useful to establish provenance. It is far easier to train distributors and security staff in cursory inspection techniques than teach them how to check for differences in manufacture or to examine label and

| PRIMARY (OVERT) | Visual and tactile security features that can be verified by the public and by inspectors, banking & retail staff. Such tests can be undertaken in no longer than 10 seconds |
|---|---|
| SECONDARY (COVERT) | Hidden (or masked) security features that can only be accessed and substantiated through the use of simple checking apparatus such as UV lamps, loupes and laser penlights |
| TERTARY (FORENSIC) | Highly complex security features that require close scientific evaluation to determine their presence and veracity. Such tests usually require some time to complete |

**Figure 2.6 -** Print and material checks can be inbuilt at a number of levels in labels and packaging to establish authenticity

| PRIMARY (OVERT) | Watermarks in paper<br>Holograms<br>Embossed close register foils<br>Color change inks<br>Security threads in paper<br>Simple codes for instant checking with an app such as NFC |
|---|---|
| SECONDARY (COVERT) | Digital watermarks<br>Micro-printing<br>Hidden image technology (HIT)<br>Complex codes and coding technology for internet connectivity via wi-fi or 4G (SMS) (QR Codes) or encrypted RFID<br>Ultra-Violet & Infra-red inks<br>Laser activating inks |
| TERTARY (FORENSIC) | Material Biometrics<br>Complex nano-fonts for microscopic reading<br>Forensic markers<br>DNA type inks for forensic analysis<br>Artificial random feature analysis<br>Spectrographic signature analysis |

**Figure 2.7 -** Illustrations of print and material checks that can be inbuilt at a number of levels in labels and packaging to establish authenticity (The above security technologies are explored in the following chapters)

packaging for minor defects in print and spelling errors that often accompany fake pharmaceutical and other products such as wines and spirits.

The presence of a security feature such as a hologram also acts a deterrent to counterfeiters, since it displays an awareness that be brand owner is investing in protection mechanisms and most likely has deployed other non-visual safety devices in order to back up any legal response that may become necessary.

It should also be noted that legal action in isolation to IP abuse is often very expensive and often comes too late in the day to deliver a proactive line of attack. Much better tactics utilize a holistic approach that incorporates a strategy that includes on-pack and in-product security features, together with inspection through the whole supply chain and invited customer participation through educating the end-user in checking techniques that establish authenticity at point of purchase.

From this point forwards, focus will be made on the ability of the print and packaging industries to provide innovative solutions to the problems related to product fraud.

# Chapter 3

---

# The importance of printing substrates in brand security

---

In Chapter 1 reference was made to the requirement for packaging and labels to protect, contain and inform as well as manage risk on behalf of the brand owner. Since materials such as metals, glass, paper, board, plastic and flexible films are the basis for all materials used in containing and protecting products, it follows that these resources offer an ideal platform on which security features such as tamper evidence and authentication systems may be based.

---

Such materials are widely available in their basic form for onward conversion in to common forms of wrappings and containers. In order to make these constituents more secure, so that they can act as indicators of provenance, it is necessary to modify them in some way so that they may be easily distinguishable and convey information that can be useful in establishing whether a pack has been opened previously, or the product labeling should be regarded as unauthentic in some way.

So that security is not compromised, it is essential that such protected materials are not available widely and that supplies are limited to bona-fide security-related printers and converters only. In order to achieve this objective, recognized and trusted suppliers of security materials and substrates will endeavor to ensure that such security materials are only made available to trustworthy converting partners within the industry.

## PROTECTION FOR NON-ORGANIC PACKAGING MATERIALS

Initially, this chapter will focus on non-organic materials that are used for packaging applications.

Metals, in the form of aluminum and tin plated steel are the primary materials we are dealing with here. Both are inherently difficult to secure (in raw material form) since they are not able to carry easily accessible markers that would be useful in assuring provenance. Therefore most security features need to be added further down the supply chain in the form of labels, inks or serial marking placed using inkjet print heads or laser engraving. More detail on this topic will be made available in later chapters.

The one notable exception to this theme is the use of holographical, decorative and security embellishment to the raw material that can be applied before it is converted into cans or box format.

Holographic designs can be applied using a laminate or varnish onto which the optical effects are engraved using high definition shims and very heavy pressure. This gives the material a diffractive finish that can be formatted to provide a continuous image that delivers similar effects to those seen on elementary holograms (see Figure 3.1).

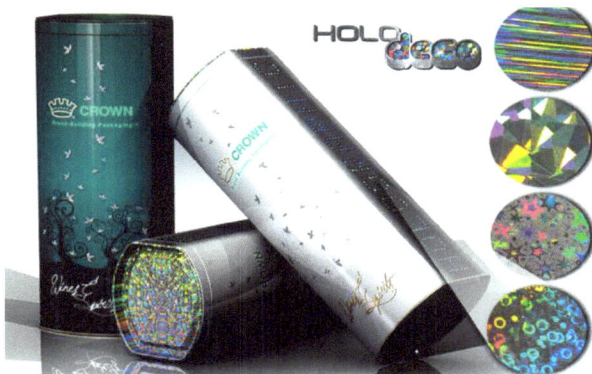

**Figure 3.1 -** Holographic effects can be added to metal based packaging in order to enhance its appearance and supply a high degree of surety that the pack is genuine

Metal is also a major component in securing metal closures on bottles and jars, mainly to seal in the contents and prevent leakage during transit but also to ensure that tamper evidence is built into the opening system so that a user is assured that the container has not been refilled or the contents tampered with.

There are various types of tamper evident metal closures, the most familiar being the 'Crown' cork. Others include a screw closure that incorporates a tamper evident band that is held in place by a perforation in the metal near the foot of the closure. When the closure is opened the perforations are broken and the released metal band can be seen to leave a visible gap between the top and base of the closure.

Because there is always a major risk of high value beverages and spirits such as whisky and brandy being attacked through refilling or dilution assault, it is also possible to add non-refillable systems to bottles and these are usually fitted in conjunction with the tamper evident metal closure.

Such non-refillable systems consist of a one way value which operates with a specifically designed pourer to stop product dripping down the neck of the bottle. Non-refillable systems are fitted in such a way that any attempt to remove them results in breakage to the bottle.

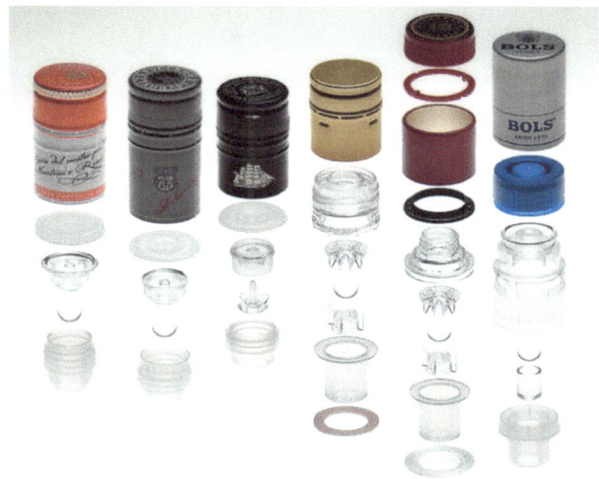

**Figure 3.2 -** Various tamper evident non-refillable systems exist. Picture: Guala Closures

It should also be remembered that metal is the most widely used material for the twist off lids used on glass jars. Tamper evidence on these components is delivered through vacuum sealing the container after filling and installing the closed twist cap. The vacuum draws down the closure and allows for a 'button' mechanism built into the closure to remain firmly in place when pressed. In this form the cap is safe as the vacuum is only released on opening. After opening the button moves up and down when pressed and provides an audible warning. A message on most closures advises to 'reject if button clicks on pressing'.

Metal, such as aluminum in thin rolled form, is also used as a lidding closure although metalized polyester type products are more popular in food and drink applications since tamper evidence is obtained through the use of heat sealing the film to the container to form a protective bond that guards against tampering and keeps the product fresh at the same time. It is possible to produce both these materials with a difficult to copy optically variable design that offers the twin benefit of tamper detection and authentication through the application of one device.

The combination of such films with blown or vacuum molded containers can also be seen in use for protecting non-food items such as printer cartridges and engine oil packaging. Both products are vulnerable to counterfeit and refilling attacks.

Finally, metal is a material that lends itself to rolling and tactile forming so that cans may be produced. Special opening mechanisms such as pull caps and the traditional use of a can opener make these components pilfer proof without further need for elaborate anti-tamper features.

The other inorganic packaging material that is difficult to protect at base level is glass. It should be recognized that glass containers can be colored and formed in many shapes and sizes and this requires a high degree of skill and expensive plant and equipment which is generally outside the reach of most counterfeiters.

For this reason most glass vessels of individual design, such as perfume bottles (Figure 3.3) and such like are relatively safe from the attention of copycat attacks. This does not mean to say that they are safe from refilling assaults though.

However, generic glass bottles such as those used in the wine industry are easily available either as in empty or previously used and discarded format, or new from a crowd of suppliers.

In these cases it is necessary to protect vessels from refilling with shrink sleeve cap seals and to add primary identification markings in the form of laser engraving or by adding further security features in the form of labels or cap protection devices, more of which will come later.

A further point worth remembering is that many injectable medications are packed in glass vials (Figure 3.4) and these too are at risk from tampering and re-filling. A wide range of tamper proof and tamper evident sealing closures are also available for 'at risk' applications such as these.

**Figure 3.3 -** Examples of individually-designed perfume bottles

**Figure 3.4 -** Injectable medicines packed in glass vials

## PROTECTION FOR ORGANIC PACKAGING AND LABELING MATERIALS

The use of organic materials such as paper, board, plastics and more recently organically derived polymers and synthetics form the most widely used materials in packaging and labeling applications.

### Paper

This material is by far the most popular for labeling applications because it is economic and can be easily transformed into labels.

Paper is easily printed and can be converted into both wet glue and pressure sensitive format in a variety of styles such as sheets, rolls and die-cut packs for onward automated application to containers, jars, cans and bottles.

Furthermore paper is endlessly flexible inasmuch as it may be manufactured in a range of colors and finishes and can be combined with other materials in order to provide decoration, instruction and of course security.

A further benefit of labels in paper format is that they are not robust enough to survive refilling attacks, since they are easily degraded and destroyed during attempts at removal, either before recycling or for re-applying to containers or bottles of fake product.

Paper is also the most trusted of materials since it is extensively used for high security print products such as passports and currency.

Therefore it should be no surprise that security labels manufactured from this resource mimic many of the safety features found in paper money and traveler identity documents.

The most predominant security feature that is used in security paper is the watermark. This feature is added to the paper at the 'wet end' of the papermaking machine and through the use of a specially constructed meshed roller the fibers in the wet wood-pulp are arranged into a form delivers a mono-tonal image that can be viewed in transmitted light.

The most commonly used process to make watermarked paper is the Fourdrinier process and the distinctive marks are formed by wires in a cylinder known as a 'dandy-roll'. Because papermaking is such a complex operation and requires high investment as well as highly skilled operators, adding

It should be noted in passing that some sources claim that synthetic label materials are more easily recycled than their paper equivalents, since synthetic labels are separated during the recycling process and recovered for use in various polyolefin compounds.

Paper labels, especially those used on glass or plastic containers break down and create a mushy pulp that has to be sent to landfill.

security features to the material is an effective method of thwarting copy attacks since these economies of manufacturing scale are not easily available to the counterfeiter.

Furthermore it is also possible to add distinctive marks, similar in appearance to watermarks at the calendaring end of the machine where mineral coating with various types of clay takes place which seal the paper ready for printing. Specialist papermakers have offered this alternative to a traditional watermarks since Fourdrinier marks are not very clear in non-transmissivity, such as when used on opaque packaging surfaces.

A further benefit of applying embedded security markings at the calender stage is that the images become even more pronounced when exposed to UV light. This feature adds a further security benefit to this alternative process.

The most secure method of watermarking paper is

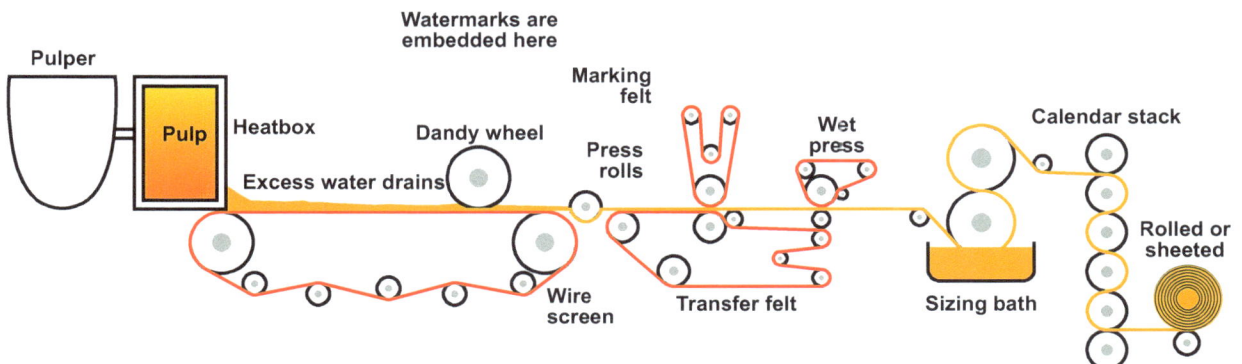

**Figure 3.5 -** Schematic of a Fourdrinier Paper making Machine

obtained by using the cylinder mold made process. This technique is considered to be more secure than the Fourdrinier process as it is much scarcer in terms of producers and provides a much finer degree of tonal detail to the watermark. This is why the process is preferred for the production of passport pages and the raw material for paper currency.

Watermarks created using the cylinder mold process are also highly visible in reflected light so they can be easily recognized when they are used on labels that will be applied on solid, opaque surfaces such as dark colored glass, metal or plastic.

Mold made paper, with accompanying watermarks, can also be lined onto pulp or paste board used for carton manufacture and subsequently utilized for pharmaceutical boxes or in cosmetic and perfumery packaging.

**Figure 3.6 -** A mold watermarked paper lined onto a pasteboard carton provides good visibility in reflected light

Watermarks may be placed in discrete register with labels so that they always appear in the same position or alternatively they can be incorporated in a continuous band in line with the machine direction of the web.

If discrete watermarks are chosen as a method of securing the material then the sheet will need to be registered with the print during the set up process on the printing press.

If a continuous web of material is being converted, as you would expect for roll label pressure sensitive production, it is necessary to fit a web guiding and register device to the printing machine in order to keep the mark in constant register with the printing web.

Established suppliers of pressure sensitive watermarked label material will be able to assist with the design and placement of a watermark on a self-adhesive substrate. Since discrete personalized watermarks can be expensive to originate and material will only be supplied in volumes high enough to set off the costs incurred with setting up a papermaking machine, it may be more practical to adopt a 'stock' design for shorter runs.

A number of alternative security features can be added to paper in order to check its provenance and make it difficult for counterfeiters to copy.

In its raw wet pulp form the material is fluid and it is possible to add visible and invisible 'markers' to the substrate at this time. Markers can take the form of colored fibers, small polyester disks (planchettes) or microscopic hi-lites. The latter are UV light reactive particles that shine like stars when irradiated with UV light.

Fibers can be added in various colors and at various lengths so that they can be observed on the surface of the substrate either by looking for them with the unaided eye or by exposing them to UV light. Measuring the length, distribution and color of these fibers allows for an individual 'fingerprint' to be embedded in the material, thus making it almost impossible to replicate by anyone wishing to copy this security feature.

It is also possible to embed narrow, two or three millimeter wide, security threads made from printed or metalized polyester in the security paper used for packaging and labeling requirements. Identical features can be viewed in banknotes and these threads can be deeply implanted in the material or made to alternate between the surface and interior of the paper in conjunction with the watermark (see Figure 9.1 on page 83). There is also potential to add further security to these threads by applying coatings that react to thermal stimulation such as body heat from a finger or from friction through rubbing the surface of the thread.

Finally, it should be recognized that optically dull paper or board should be chosen for security applications where invisible UV security inks are to be applied during the printing process. This advice is applicable to inks that react to UV light and should not be confused with UV ink curing systems which is an entirely different process and used for drying inks and not for authentication purposes.

It will be discovered later in this training module, security paper used in conjunction with other complementary security devices, can be a resilient defense against both copy and tamper assaults.

### Board

Carton board for security applications consists of both pasteboard and pulp board. A slightly different approach is required when adding security features to these materials. Together, these substrates provide an excellent primary packaging material as well as offering a handy platform for important supply chain information such as route to market, manufacturing source and expiry data to be added.

Board products are also an ideal material for swing and hang tickets that can be used to provide useful guidance on a product's capabilities, contents and size. In swing ticket and hang tag format both products can carry their own individual security features or be further embellished with security inks, foils or RFID labels.

Paste board, as its name implies, consists of a number of paper plies which are bonded together to form a thicker sheet that can be used to make boxes or cartons that protect products in transit and in storage before use.

These individual paper plies can carry their own authentication features in the form of forensic markers or more cost effectively offer a low grade security feature through mixing colors within each ply so that a 'sandwich' of colored material is created. Authentication is simply a matter of tearing the tag to reveal a colored core.

It is also possible to metalize board products so that they display a variety of colored shades and also

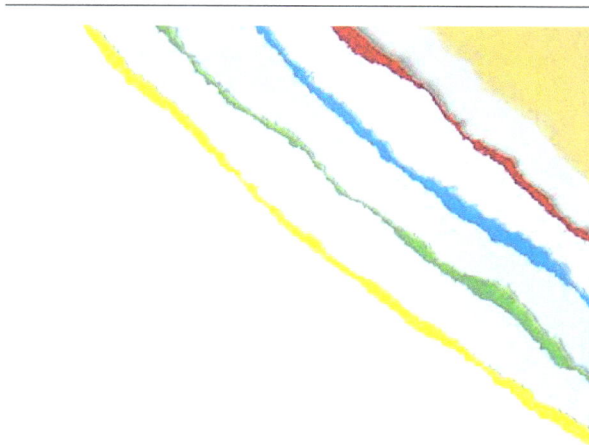

**Figure 3.7 -** Various colors are available to provide a quick 'tear and check' feature to check the provenance of a paste board swing ticket or tag

to emboss the metal coating to deliver an optically variable image or holographic decoration, both of which offer certain low grade security devices by making the material difficult to replicate by scanning or by attempting to copy the design using publishing software and a desk top printer.

Other fraud deterrence techniques include surface embossing to create discrete patterns and film coating which delivers various optical effects.

### Material biometrics

At this point is worthwhile exploring some new and interesting technologies that utilize the surface properties of material at a nano-level to deliver an assurance that the material is original and not a fake.

Whilst these technologies are not per se print related they can be used to authenticate the very material on which the print resides. This may be a label, a carton or a plastic container.

At a microscopic level each square millimeter of material is different. As an illustration of this fact a piece of paper is composed of millions of discrete fibers, all interlocked to form a continuous sheet.

If it was possible to 'grab' an image of say a specific square millimeter of the surface of a label –

each label's top right hand corner for instance – it would be observed at a nano-surface level that each was different and carried its own individual mat of fibers arranged in a dissimilar pattern.

Technology now exists that can record a 'digital fingerprint' of a predetermined, minuscule piece of material on every label, making it possible to identify the provenance of each one accurately. The process is also referred to as random feature identification since it allows items to be identified through the random changes that each item possesses at a microscopic level. Sometimes these may be imperceptible changes in a letter of type – which will be addressed later – or imperfections in the material surface.

In biometric terms this technique is similar to the fingerprint recognition methods used by the police to identify suspected felons at a scene of crime. A search for one fingerprint in a database of many millions takes time but this can be shortened through the use of complex software and high power computers.

Material biometrics works in a similar way but can deliver a positive or negative result much more quickly if the 'fingerprint' from the material is linked to a sequential barcode. This allows the barcode to act as an identifier and as a link to the surface material scan and its corresponding record in the database.

Various other methods exist that include the recognition of visible fibers randomly distributed in the material mat and laser illumination of the material's surface and analysis of the reflected light from each predetermined area of a label. For speedy recognition both these methods also rely upon a barcode or other machine readable reference from which to match the image. Further detail on this subject follows in a later chapter.

This whole fascinating field of material biometrics is changing constantly as imaging and computer processing and data management skills progress. Technology enablers such as cloud computing, smart phone apps and scanners as well as 4G communication systems are all conspiring to provide more on-the-spot identification methods for both the police and those investigators involved in brand protection.

## Plastics and synthetics

These materials are very hard to protect from counterfeit attack. The best method of defense is to follow similar procedures to those used in glass containers; stylish, registered designs that are specific to each product should be considered since having to copy discrete designs and molded inlays can be restrictive when it comes to try to knock-off a particular brand.

This technique is not sophisticated enough to deter the determined counterfeiter though.

Further protection can be achieved through the selective use of micro-taggants which are uniquely identifiable particles embedded in base material and identifiable through proprietary readers or laboratory analysis. More explanation on this technology will be explored in a later chapter.

## Metalized films

Metalization adds a layer of complexity to films that can act as a barrier to counterfeit attack. Again this process is unlikely to discourage the determined counterfeiter. Opportunist attackers however will find it off-putting and may well try to experiment with an alternative target that takes less effort.

Metalization is mainly used as a barrier coating but also for decoration. In decorative form as will be discovered later, can be useful as a base for creating difficult to copy optically variable features such as holograms and high resolution multi-diffractive effects that are resistant to copy attacks and counterfeiting in general.

## Tamper evident adhesives and tapes

Common usage in the labeling and packaging industry of the terms tamper evident, tamper resistant and tamper proof understandably lead to a good deal of confusion.

This is because these terms are used to describe a variety of similar functions that are designed to combat some very different threats or risks.

Tampering with labels to remove or mask fraudulent activity has been a threat ever since the introduction of self-adhesives over a half century ago.

The benefits that are available, for instance on

removal of price marking labels, allow perpetrators to exchange the labels on low cost items for those carrying a higher price. This is why the majority of price marking labels carry distinctive cuts and indentations around their circumference. Attempted removal leads to destruction of the label and immediate evidence that an attack has occurred.

Such benefits may be viewed as negligible now since price marking labels have all but disappeared with the growth of universal product coding and item scanning at check outs. Nevertheless this acts as an easily recognisable use of security cuts on a self-adhesive label.

### Trends in tamper evidence

Tamper evidence has since evolved into a number of distinct forms, all aimed at protecting products from unwanted and often dangerous activity such as refilling, product spiking – which is linked to extortion attacks – and pilfering which involves taking part of the contents out of a sealed pack and then resealing it to conceal the fraud.

**Figure 3.8 -** Various methods of introducing tamper evidence to a container through the use of alternative label designs

Primarily the technology provides visual confirmation that a product that has been sealed has not been previously opened. This is achieved through the application of tape or a label over the closure or vulnerable points in the pack or at point of contact between the lid and body of the container.

However, as such tamper evident indicating products have evolved so have the techniques to compromise them.

Sophisticated stress-indicating materials and adhesives that are resistant to temperature changes are now being combined with authentication devices

in an attempt to combine these essential functions.

The pressure sensitive materials industry has developed a range of 'VOID' substrates that separate when removal is attempted and the word 'VOID' appears when the top of the label is peeled away after fixing down. Otherwise a bespoke approach can be taken using a brand name that becomes tamper evident when opening a carton holding valuable goods such as a computer or smart phone.

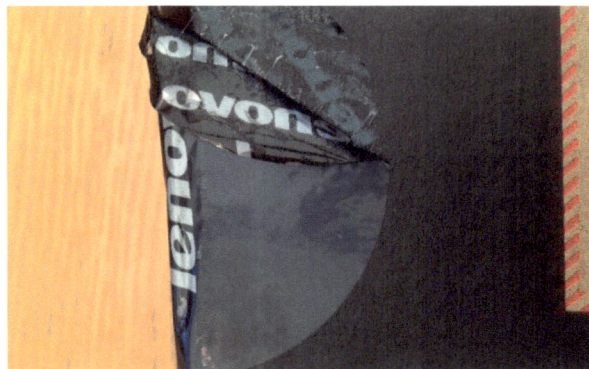

**Figure 3.9 -** A brand name is revealed on opening this tamper evident closure

Other methods involve the use of specially selected traditional paper face stock with good resistance to tear. It is necessary to also have a good understanding of the packaging material that the tamper evident label will be applied to. Therefore it is often necessary to work closely with an adhesive supplier and the pressure sensitive adhesive supplier in order to achieve the best results of 'initial tack' and permanence. This is especially important where labels are being applied during an automated container filling process.

An alternative approach is to use 'destructible vinyl' face material for the tamper evident label construction. Vinyl face materials are high performance filmic products that are frangible on removal. Such products are utilized in heavy duty applications where resistance to moisture, heat, cold, dirt and grease would interfere or degrade paper.

Evidence of tampering is provided by the material

itself fracturing into minute pieces as soon as removal is attempted. As removal involves a certain degree of 'picking' at the edges of the label the material fracturing becomes evident thus visually drawing attention to the attempts. Further security can be added at no cost by printing a solid colored border around the edge of the label making tampering even more evident.

By adding holograms plus track and trace technology, and in some cases RFID, these tamper evident devices will become an important component in future brand protection applications where anti-theft properties as well as protection from refilling and counterfeiting are important attributes (Figure 3.10).

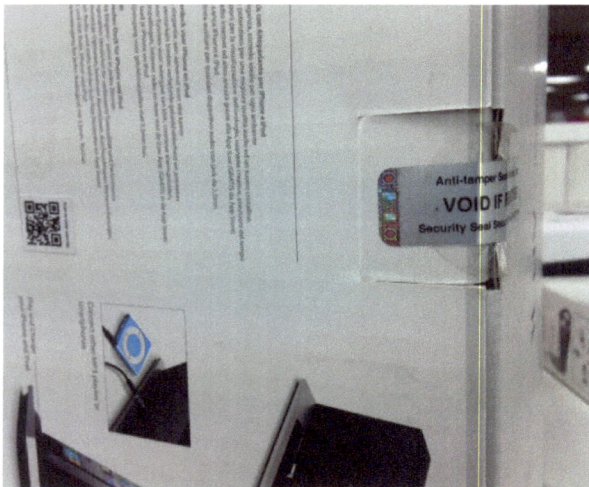

**Figure 3.10 -** Tamper evident devices become an important component in brand protection applications

Sophisticated shrink wrapping films aided by specially marked tear tape and colour changing effects are now being combined with coding technology to protect bottles and similar containers from refilling, counterfeiting and dilution attacks. These take the form of shrink sleeves that are applied over the closure in order to provide protection and indication of first opening.

Developments in disabling RFID tags in an attempt to 'kill' them after a product is opened to prevent

re-use also offers a dual tamper protection benefit to users.

Material science continues to make headway in this important area with a number of universities in the US, Europe and China now involved in researching and developing films and adhesives to meet the threats of the future.

**Shape memory polymers**
One further technology that is beginning to set foot in this market is SMP (Shape Memory Polymer). The material changes shape and form at specific temperatures thus providing an indication of authenticity and temperature change in the supply chain.

SMP materials can be converted into labels and tags and therefore provide a new range of opportunities for the brand protection industry as a whole.

Shape Memory Polymers are smart materials that have the ability to return to their original shape from a temporary state through the introduction of a trigger event – in this case heat.

Such label materials can be constructed to 'store' 3D shapes in the form of embossed text or logo's and then programmed to release these when exposed to a specific temperature change, thereby creating unique massages that can be utilised as brand protection confirmations or specific safety alerts.

**Figure 3.11 -** Shaped memory polymers provide an authenticity check on neck sleeves and hang tags

The secret behind such materials lies within their molecular structure and their ability to change state rapidly when exposed to the correct trigger – in the case of the materials illustrated above this is 65°c.

Whilst some work is being done on triple shape memory materials that change shape and then revert back to their original shape after the trigger event, most SMP materials are irreversible after activation.

Since there is no need for specialist activation/ authentication equipment the materials are ideal for consumer verification since all that is required to set the change in motion is a cigarette lighter or hair dryer.

It is too early to appraise the attributes of such a material yet but it shows promise in tamper protection when combined with the functions mentioned previously.

### Forensic markers and taggants

Anyone who watches popular crime scene investigation TV and direct streaming programs will be familiar with forensic analysis that relies upon the principle that where we go we all leave microscopic traces of our passing along the way. This may be in a trace of DNA left on a cup or glass or a fiber from our clothes. These traces can be used in evidence should any of our misdemeanors be challenged or brought before a court of law.

Forensic markers and taggants are unique artificial and natural compounds that can be placed in suspension within a liquid or coated on to a solid object and even incorporated, at infinitesimal levels in the material used to make packaging and labeling substrates such as inks, paper, plastics and also, adhesives.

These substances consist of naturally available, but rare compounds that are mineral based or artificial organically produced ingredients that mimic DNA in their construction.

This topic will be revisited in more detail later in this study but at this time it is sufficient to know that many products today carry these forensic markers as a definitive measure in order to prove provenance and also purity.

In defining provenance we are referring to authenticity and a check against unauthorized copies being made. In terms of purity it should be recognized that where there is a risk of dilution attack, then covertly added ingredients that can be recognized in proportion to their mix, within a liquid or powder, are useful devices to measure unsanctioned dilution.

This measure is obtained through minute levels of taggant that are placed in (say) a liquid at a ratio of one part per billion. This is a quantitative amount so it should be present at all stages of the distribution system. If later analysis provides data that shows (say) one part per ten billion then it is provable that unauthorized dilution has occurred.

Likewise forensic taggants can be placed in inks, paper or board and then measured and recognized downstream in the distribution cycle in order to detect counterfeit products that may have infiltrated the supply chain.

# Chapter 4

# Inks, coatings and varnishes – safety explored

Ink is, of course, an indispensable component of any label or packaging application. Without ink it is not possible to reveal what is inside a pack, instructions regarding usage and of course who made the product. A blank label is essentially useless.

Since ink is used to open a visual communications channel between every party in the process of product distribution and final use it follows that it can also offer a handy method of communicating provenance at the same time.

Of all the security related print devices available for brand protection and securing packaging and labeling from threats of counterfeiting, tampering and alteration of important information such as sell-by dates, identification references and variable supply chain data, inks are the most versatile component in the armory of the printer and converter.

Regardless of the printing process, whether it is litho, letterpress, flexo, gravure or silk screen, there is a whole range of inks available to act as precursors of authenticity, indicators of alteration and covert messengers that can be used to conceal hidden information that is only accessible to those with the knowledge and equipment to reveal their secrets.

In itself, ink is a versatile commodity that can be adapted to carry security features that are sensitive enough to provide reactivity to external stimuli such as light and heat. If used correctly, these responses deliver an indication that a pack can be trusted or

that a label is not all it purports to be.

In order to protect security inks from unauthorized use, it is necessary for the industry to establish that it is dealing with bona-fide printers that can be trusted before supplying product to any printer or converter. This is a necessary prerequisite of any of the materials utilized in product protection and applies to paper, board and foils as well. Safeguarding such supplies whilst they are in stock or on the print shop floor awaiting conversion is an important responsibility for the printer too.

Since there is a tendency for packaging and labeling converters to generate product using the four color printing process it is necessary to point out that in order to provide security it is obligatory to deliver a security ink using spot color and this requires extra print stations to be available depending upon the number of colors needed in excess of cyan, magenta, yellow and black (CMYK).

This in itself is not a hindrance for most suppliers since they will already have equipment capable of delivering a number of additional colors over and above CMYK, but for those with a basic four color limit, or those with only digital print facilities, security

```
┌─────────────────────────────────┐
│     Inks for primary validation   │
└─────────────────────────────────┘
```

| Optically variable | Iridescent | Thermo-chromic | Coin reactive | Penetrating |

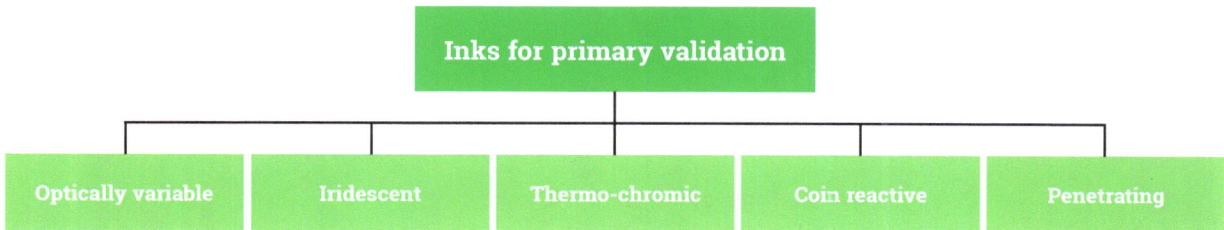

work will require additional runs through the press or the utilization of alternative security devices if only digital methods of production are available.

As hinted at earlier, there is a wide choice of security inks, all designed to deliver specific reactions in order to safeguard against a whole range of potential threats. These threats range from unauthorized replication right through to detecting alteration and substitution attacks.

Instances include printed codes that may be erased or replaced in order to disguise an out of date product, or to mislead consumers by indicating that a component is suitable for a particular use when it is not.

An illustration of 'change of use' would be where a code on an electrical product label specified it was safe for industrial use when it had only been authorized for domestic applications. Such changes would immediately increase the 'value' of a product by just changing or altering the identification code on the label.

Before deciding on the suitability of ink as a method of adding a security feature to print it will be necessary to evaluate the protection from potential risks that such a solution will provide. Decisions should be made relating to whether protection is required at point of sale or right through the supply chain.

Additionally, it will be essential to establish whether an ink will be used as a primary, secondary or forensic identification feature, or maybe all three. Designers should appreciate that if they are also to incorporate supplementary devices such as holograms or serialized codes then consideration must be given to how these all complement each other in the system and whether any unnecessary duplication of function is encountered.

It is, for instance, a duplication of function if a hologram is used for primary recognition and combined with another optically based feature such as a visible color change ink. For low to medium type security protection, such combinations can add considerably to the cost of the final product with little additional benefit.

There follows an inventory of inks and vanishes that are suitable for security packaging and labeling applications. This list is non-exhaustive since new inks and pigments for such uses are under continual modification and development.

## INKS THAT MAY BE USED FOR PRIMARY VALIDATION PURPOSES

These inks are designed to supply an initial indication of provenance that may be obtained visually within a few seconds.

### Clear or optically variable varnishes

The use of varnish in print provides a matt/gloss effect that is highly resistant to scanning and attack through the use of 'home office' laser and ink jet printers that are often used for low end replication of counterfeit labels and packaging such as small cartons.

By adding a color shifting pigment to the varnish security can be increased even further. In the illustration (Figure 4.1) it should be noticed that braille is also present on the carton, as a prerequisite of pharmaceutical requirements for those that require tactile confirmation on the pack as they may be visually impaired.

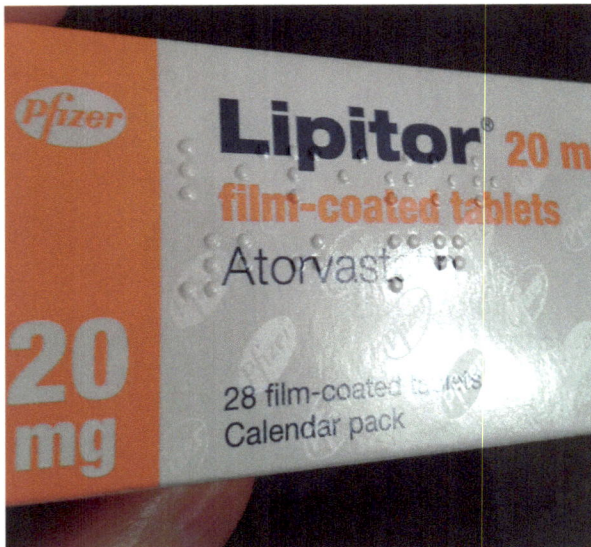

**Figure 4.1 -** The use of a varnish seen here used as a logo appears in positive/negative form when tilted to light

**Figure 4.2 -** Optically variable inks, similar to those on banknotes change color significantly when titled

Because of the size of the color shifting pigment particles in the ink it may not always be possible to use some printing techniques such as litho and flexo to deliver optically variable varnishes on some materials.

### Optically Variable inks

As their name implies, optically variable inks (OVI's), visibly change color when tilted by the observer to deliver an easily recognizable shift in appearance.

The color change is delivered by millions of small light reflecting platelets distributed within the ink. These shiny substances provide a very definite change in color when tilted and are widely used in product protection as can be seen from Figure 4.2.

Such inks were originally formulated for banknote protection and understandably they are under continual refinement and development. Basic OVI's for brand protection are restricted to specialized suppliers and only the highest security products are used in currency applications.

Because such inks carry high levels of pigmentation to deliver their optically shifting effects it

may be necessary to use printing processes that are designed to carry heavy ink weights such as silk screen. A high degree of consultation with the ink manufacturer is recommended before OVI work is undertaken.

### Iridescent inks

Similar in visual results to optically variable varnishes, iridescent inks deliver a multitude of colors in an effect that is similar to that observed on bird's feathers or the wings of a butterfly.

Such inks are widely available and used predominantly to decorate cosmetic and body care products. Unless an ink of this type is specially developed to provide a unique visual appearance it should only be considered for decoration and very low level protection applications.

### Thermo-chromic inks

These products react to specific variations in heat and are mainly used to temporarily indicate a certain degree of chill has been reached before a beverage such as beer is consumed, or that conversely a warm drink such as coffee is still too hot to consume. Communication is achieved by a brief color change from clear to deeper shades of blue in the case of chilling and towards orange and red in the event of a heat warning.

Thermo-chromic products are also useful where permanent records of temperature exposure are required. In these instances color changes are designed to point out that a product pack or label has been exposed too long to a specific temperature threshold that makes the product unsafe or unusable. In medical applications it is useful as a tool to measure correct autoclaving temperatures have been reached.

In product security applications, thermo-chromic inks are used as an indication of genuineness since they can be formulated to react to body heat in the form of pressure from a thumb or finger and change from color to clear or the other way round from clear to color. Alternatively such inks can also react to heat created by the friction generated by scratching the ink with a rough object such as a coin or a thumbnail. (note: some ink encapsulation processes deliver a similar result)

It is also possible to embed thermo-chromic properties into plastic containers and closures which enhances their use even further.

WARM

COLD

**Figure 4.3 -** Some color reactions to changes in temperature are shown

Recent developments have delivered inks capable of indicating a variety of color changes depending on the temperature they are exposed to. These are referred to as tri-thermochromic inks.

### Coin reactive inks
A very basic level of security can be achieved by the use of coin reactive inks. These products are visible as a semi-gloss varnish after printing and when a coin is rubbed across the ink a reaction occurs, causing minute amounts of surface material caused by oxidation on the coin's surface to be transferred to the clear ink creating a gray color.

### Penetrating inks
Again, as their name implies these products are designed to infiltrate paper substrates so that they bleed through the material and can be easily viewed from the reverse. Generally these inks appear black on the exterior of the paper and show as a color on the opposite side. Well performing penetrating inks will be absorbed into the matte of the paper and will be very difficult to erase or alter. Therefore they are useful for delivering secure product coding that may be at risk from alteration or erasure.

### INKS THAT MAY BE USED FOR COVERT AUTHENTICATION APPLICATIONS
We should remember that early in this study we discovered that if there was any uncertainty regarding the provenance of a primary or overt (obvious) security feature, then some secondary degree of confirmation was required in order to make a sound judgement about authenticity.

These 'secondary' procedures require an instrument of some sort that can be used as an assurance that a security feature is present – or not.

### INKS THAT ARE RESPONSIVE TO ULTRA VIOLET LIGHT
Over the years, one popular method of achieving this secondary authentication objective is to use non-visible light in the form of ultra violet energy delivered by a power cell or mains connected lamp. These are also known as 'black lights' and need to be shaded from direct visible light in order to work efficiently so

```
                    ┌─────────────────────────────────────────┐
                    │     Inks used for covert authentication   │
                    └─────────────────────────────────────────┘
```

| UV responsive | Metameric | Photo-chromic | Conductive | Machine readable | Tamper deterrence | Forensic |

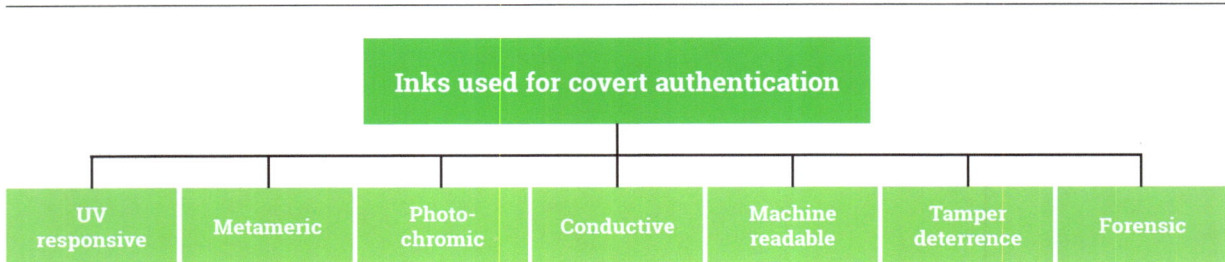

that UV reactive print may be observed.

Inks can be formulated to respond to both short wave and long wave UV light. However, since long wave UV activating inks have been present for a number of years they are not considered as secure as their short wave alternatives because the chemicals used in short wave UV inks are not so easily obtained and in some countries are strictly controlled.

More recently, UV inks have been modified to deliver much more secure responses to UV light and it is possible to 'tune' such inks to react to much narrower wavebands of light at both ends of the UV spectrum.

There is a wide range of colors available and it is not necessary to adopt the traditional blue hues that have previously been used and can be easily

compromised. Indeed there are products that turn from visible to invisible and those that offer a distinctive color change through three of four variations of shade when exposed to different intensities of UV exposure.

Ultra violet reactions are a form of photoluminescence. But this reaction ceases the moment the UV light source is switched off.

An additional reaction, known as phosphorescence can be made to deliver luminescence that continues to fluoresce after the UV light source has been extinguished. The intensity of the luminescence displayed by the printed image then decays over a period of time that it can be measured accurately. For this reason such inks are very useful to security printers because they can be tuned in both

## The Electromagnetic Spectrum
### WAVELENGTH IN NANOMETERS (MILLIGRAMS)

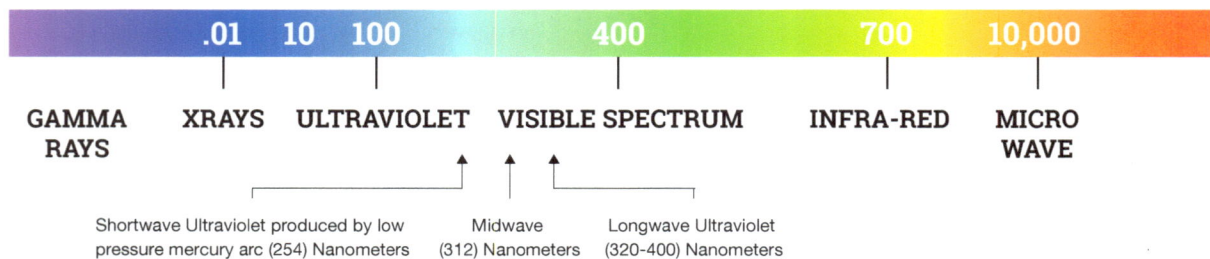

| .01 | 10 | 100 | 400 | 700 | 10,000 |

| GAMMA RAYS | XRAYS | ULTRAVIOLET | VISIBLE SPECTRUM | INFRA-RED | MICRO WAVE |

Shortwave Ultraviolet produced by low pressure mercury arc (254) Nanometers

Midwave (312) Nanometers

Longwave Ultraviolet (320-400) Nanometers

When properly filtered, Longwave or UVA ultraviolet includes radiations that lie just below the visible spectrum in the range of about 315-400mm. Midwave or UVB ultraviolet lies in the range of 280-315mm. Shortwave or UVC ultraviolet is 257mm.

**Figure 4.4 -** The electromagnetic spectrum and the infra-red and ultra-violet wavelengths that can be utilized for security ink authentication systems

color and the period of time they take to decay. These properties make ideal authentication devices.

## Metameric inks

Metamerism is a phenomenon that is used to describe color changes that occur in certain specially formulated inks when they are viewed under different light sources. For instance, under natural daylight a pair of color-matched inks will appear to be exactly the same. Expose the inks to an alternative light source and a very different image is observed.

Therefore metameric inks need to be 'paired' to work effectively. A metameric pair can be alternatively defined as two colors with different spectral compositions that generate the same color stimuli under certain conditions such as lighting, size and angle of viewing or the chromatic sensitivity of observers. In fact, we talk about a metameric pair because this effect is evident when comparing at least two color samples.

Such inks are available in a variety of combinations and since it is not always evident that such inks are metamerically paired on a piece of secure print it is possible to deliver a highly covert authentication feature that can act as an effective counterfeit detection measure.

## Photochromic inks

Photochromism is the ability of a chemical to respond to light and display this response in the form of a color change. In the case of sunglasses, these chemicals react to light intensity by changing to a darker shade as the sun gets brighter.

The security-related print industry has long toyed with such reactions and tested these compounds on a number of occasions. When carried in inks, the photochromic chemicals (spiropyrans) can so far only be made to react to high intensity light such as a camera flash, and further limitations are that they are affected by daylight which causes loss of color fastness over time.

The color change reactions of photochromic inks are reversible and once a change develops it lasts much longer than the changes that are observed when phosphorescent light excitation is removed.

Work is currently underway at a university in the UK to improve the range of colors available and the results so far promise an early solution to these problems.

If successful, expect photochromic reactive inks to be more widely used in future.

## Conductive inks

These products are widely used in the printed electronics market which is way beyond the terms of reference for this study.

However certain developments in this area are of interest to product security inasmuch as work carried out recently in Germany has delivered a conductive ink that can interface with the screen of a smartphone.

An invisible printed pattern is applied to the tag, label or board used in packaging for promotional purposes and authentication. This pattern acts as a 'pointer' or trigger so that when the surface of the printed item is touched to the smartphone or tablet screen it interferes with the capacitive touch functions and acts in the same way as a finger is used to tap or navigate the system.

**Figure 4.5 -** Inks that react to the capacitive screens found on smartphones and tablets can be used to trigger authentication messages

Codes can vary and when activated can 'virtually deliver' the user to a webpage where product authentication can take place. They need to work in conjunction with an app which needs to be installed before touching the print to the tablet or smartphone screen.

### Machine readable inks

In order to remove any ambiguity that may exist when authenticating inks and deciding their genuineness manually, a number of automated technologies now exist in the form of small hand held gadgets that recognize specific chemical signatures and respond with an audio or visual signal that confirms or rejects provenance of the ink. Unique chemical signatures are formulated from a number of 'rare earth' materials that are then embedded in minute amounts within a printing ink or varnish.

The particles, which are invisible to the naked eye, glow brightly when lit up with specific frequencies of light. These specks can easily be manufactured and integrated into a variety of crystallized materials, and can withstand extreme temperatures, sun exposure, and heavy wear.

These crystals are doped with elements such as ytterbium, gadolinium, erbium, and thulium, which then emit visible colors under say, near-infrared light. By altering the ratios of these elements, it is possible to tune the crystals to emit a number of colors in the visible spectrum.

Such materials are then engineered to deliver highly specific reactions that can be used in millions of disparate applications, thus protecting the integrity of each system from reverse engineering.

Readers that recognize these signatures, range from small inexpensive hand held devices that are powered from a battery, through to high speed automated machinery that is utilized to validate batches of banknotes before they are placed back into circulation after being received over the counter.

### INKS THAT ARE DESIGNED TO DETECT AND DETER TAMPERING

Tamper evident inks have been used for decades to protect financial documents from alteration and erasure attacks. They have also found wide use in the construction of tamper evident labels so that messages can be displayed such as 'opened' and 'void' to reveal that a label used as a closure can

display a warning message and be destroyed during the unfastening process.

More recently, with the popularity of product coding it has become necessary to protect products that carry date and product coding, serialization and batch specific data from alteration and/or erasure assaults.

This is because there is hidden value in these codes as they are used to identify when products should be withdrawn from sale because they have reached their 'use by' date. There is also value in protecting the codes used to identify the performance of products and there is money to be made in remarking those with low end specifications in order to disguise them as offering higher functionality. An illustration of this point is the remarking of computer processor components or graphic cards so that they pass off as more highly priced articles.

Inks used for coding then, need to be permanent and they also require a secure platform or base on which to reside. Placing a screen of erasable ink under the code offers a degree of protection, which can be enhanced further by including a degree of chemical sensitivity in the ink to protect against solvent or bleach attacks that are aimed at erasing the code so that new data may be placed down or the original code altered to deliver a different message.

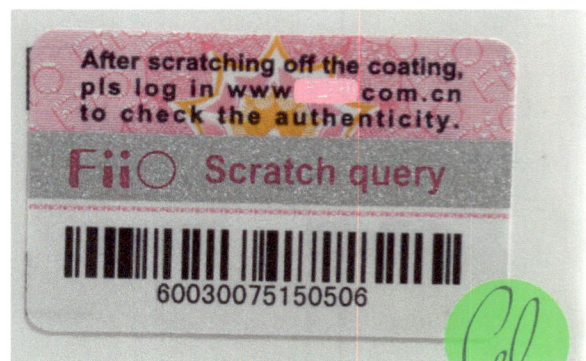

**Figure 4.6 -** Scratch off inks can be used to deter tampering especially where coded messages are used for internet checking of authenticity

Likewise, codes that are used for product authentication are often protected by a layer of scratch off ink that is removed during the verification process, ensuring that the same code cannot be reused again later.

For additional security the scratch off panel can be overprinted with a tamper evident design.

### Forensic inks and taggants

For very high security applications it is necessary to add a form of forensic security to the ink. Marking an ink or other print related material such as paper or label adhesive with a forensic signature requires the addition of a tag or taggant.

These tags are microscopic particles that are embedded in the material and are distinguishable either through their chemical or biological signature or through high power magnification. They are sometimers referred to as nano-markers.

**Figure 4.7 -** When a reader is in close contact with a taggant carried in an ink then a signal in the form of a green light is delivered

In the same way that DNA can be used to recognize and prove the participation of a criminal in a crime, taggants can be used to prove conclusively the provenance of ink or label material on a label or package.

Indeed, biological materials in the form of synthetic DNA are used in product protection programs. The presence of the DNA is identified either by laboratory testing which takes some time, or through the use of specially doped chemical testing strips that work in a similar way to pregnancy testing. Color changes on the strip indicate the presence or absence of the validating DNA tag.

To simplify the process of categorizing taggants further it should be noted that there are two classes of nano-marker, organic and inorganic.

Organic markers such as synthetic DNA are easily dissolvable and can be carried in solutions that are capable of being applied by inkjet coders. Inorganic markers are based on chemistry and the physical properties of small indicators that vary from the size of, say, a grain of salt through to material at micron levels. These particles are best carried in viscous inks or varnishes, since the process of reducing their size in order that they can be delivered through inkjet heads is a further specialized process.

At the top end of the scale, micro markers can be identified with magnification and tools such as loupes and microscopes are used to identify particles that can be visually coded with a variety of patterns, colors and stripes.

These coded particles be distributed randomly in an ink or varnish and when visually confirmed that they are present authentication is confirmed.

At a nano-level though, specialist readers are required to identify the tag and the signature carried by the markers. Signatures can range from those confirmed on a spectroscope through to specific chemicals that may be recognized by a hand held reader that displays a confirmation that the tag is present through the display of a red or green light.

Since the chemicals used in these inorganic authentication systems are based on materials that are distributed in very small amounts such as parts per billion, it is very difficult for anyone to reverse engineer the process, even if they have access to a reader and the chemical tag involved in the authentication procedure.

A further benefit of these tags is that since they are present at measurable levels, any variation in the amount of tag present in a sample can be readily recognized and can act as an indicator of dilution in vulnerable products such as liquids and powders.

It is a worthy practice to embed the same markers

in the packaging as well as the product in order to obtain a systemized approach to product protection.

For instance it is a recognized practice for some producers of valuable vintages of wine to embed the DNA from their product in the ink that is used to print their wine bottle labels.

# Chapter 5

─────────

# Building the foundations – origination

─────────

Every one of us is familiar with the functionality of security print design since we come into contact with banknotes and other high security print items on a regular basis. In the arena of labels and packaging however, security design is a relatively new concept to most suppliers in this field, as they are more familiar with designing for physical product protection, together with other functional requirements such as maximum brand recognition and the conveyance of useful information regarding product usage and ingredients.

─────────

As has been explained before, labels and packaging provide an ideal platform for the carriage of indicators of provenance as well as tamper evidence and track & trace technology.

Since the space on a product label or packaging is at a premium, it follows that any additional information that may be added to improve authenticity checks and usefully provide supplementary facts such as supply chain channels and product codes should be as functional as possible. This is of course without such data impacting on, or detracting from the primary functionality emphasized above.

It should also be borne in mind that graphical design can be adapted to carry overt, covert and forensically detectable features in a similar manner to those carried in raw materials such as inks and substrates in the form of paper or board.

At this juncture it is necessary to point out that graphical design is constrained firstly by the printing methodology employed by the converter and secondly by the number of colors available to the print process being applied.

In the fast moving consumer goods sectors, which include food and drink products, color is an essential part of communication between brand owners and their consumers. Here then, process color predominates with additional spot colors and often metallic inks being deployed to decorate and attract the buyer's eye.

With industrial goods and mainstream pharmaceuticals, process color often gives way to graphics that are carried in monotones or duotones, and here a different approach to combining security features with graphic images is required.

Embossing and foiling are also important tools in graphical design and can provide useful security features as well as enhancing the visual appearance of a label or box.

Employing graphical design as a tool to deter and detect product related crime is one of the most economical methods of prevention and detection of fake products, since images and text are key components of brand recognition and information conveyance and are present wherever labels or packaging are deployed.

Whilst adding a security ink or a hologram

increases the cost of the pack or label incrementally, design is an unrepeatable charge once created and therefore a highly economic method of authenticity check that is available to the most auspicious of brand owners.

It is of course important to recognize before entering into the design process the objectives you wish to achieve at the conclusion of the project. If there are no perceived product threats, such as tampering or counterfeiting that require countermeasures to be considered then there is little point in developing the security approach further.

Of course, if there is a history of previous counterfeit product attacks or a vulnerability to refilling a pack or bottle with fake product then precautions will be required.

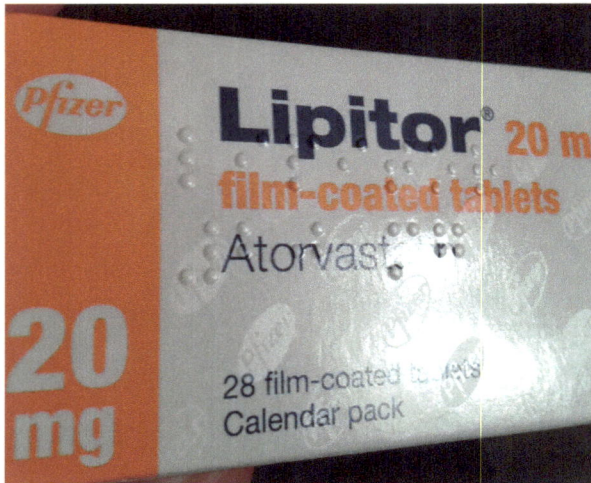

**Figure 5.1 -** A simple but effective low level anti-copy feature can be delivered to artwork by using an overprinted varnish carrying a logo in negative and positive format

The design stage of any overhauling of the packaging or labeling will then require a degree of evaluation of the threats faced and how to counter these with specific features integrated into the design and production process. This is because production is crucially linked to design and not every printing process is suitable for the deployment of all the

various procedures available to embed the wide range of security features available within a label or pack.

To provide an illustration of this constraint, not all print processes are capable of carrying enough ink weight to provide an effective color change when using optically variable inks. Other processes are not able to deliver the very fine detail that is required in order to embed covert images that can be recognized easily with high magnification and other viewer assisted processes such as smart phone cameras.

More importantly, the move to digital and hybrid printing systems that incorporate inkjet and other variable non-impact processes with conventional printing will require a different approach to design and production methodologies.

The message here then is clear. It is essential to recognize that you research all the relevant facts not only about the materials and print processes involved in delivering an effective design, but also how those procedures will deliver a satisfactory result and provide the brand with a solution that is able to identify as well as deter unwanted criminal attention or fraudulent activity.

## DESIGNING TO DETER AND DETECT COPIER AND SCANNER ATTACKS

Today scanners are ubiquitous, with high resolution devices widely available in the home office as well as the workplace.

Previously, scanning attacks designed to copy the existing labels or packaging on products under attack would only be possible in the printer's origination studio.

Nowadays, most people have access to not only a scanner but also an embedded printer so it even becomes unnecessary to move the scanned image into a publishing program to print it out.

Labels and tags are most at risk from this threat as they are produced on relatively lightweight materials but carton packs too can be copied as many desk top printer/scanners can accommodate paste and pulp-board in their feed mechanisms also.

What needs to be remembered is that scanners can only reproduce material by converting the image to a dot screen and then replicating the dots to recreate an effective copy. Dot size here is important and we refer to this as 'resolution' or dots per inch (dpi).

During the scanning process, any color present within the image will be converted to cyan, magenta, yellow and black (CMYK). As we should know, these are the basic colors required in process printing in order to reproduce an image using the conventional printing color separation process. [There are of course variations to this such as red, green and blue (RGB) used mainly in TV's, and commercial proofing systems that use hi-resolution inkjet printing that involves extra colors, but let's ignore that for the moment].

The higher the resolution of a scan, the more dots per inch are required to reproduce an image and therefore the image reproduced attains a better quality.

An understanding of this basic constraint of all scanners provides us with a variety of responses that are available to deter and detect direct copy attacks.

Firstly, the use of continuous lines in the origination process will provide a detection device since any lines copied onto the fake packaging will consist of dots, which can then be easily identified with a loupe or other magnification tool.

Using 'spot' color is a further refinement since any scanner attack will convert the spot color into its relevant 'process colors' of CMYK and be identifiable again using magnification. Spot colors that are impossible to scan such as day-glow inks, metallic inks and fluorescents are also useful and act as a further deterrent if they can be incorporated within the design artwork.

## USING SCREEN ORIGINATION TECHNOLOGY TO DEFEAT COPY ATTACKS

Screen modulation tools are also useful since by controlling and varying the dot size carried by the screen it is possible to leverage on the weaknesses of scanning systems that work on fixed resolutions of say hundredth's and thousandths of an inch. By finely tuning dot and line screen origination outside of these scanner threshold's it is possible to create hidden images within the artwork that are only visible when a label or carton is scanned dishonestly. The hidden image is then highlighted in clear view to deliver a visible message such as 'VOID' or 'COPY'.

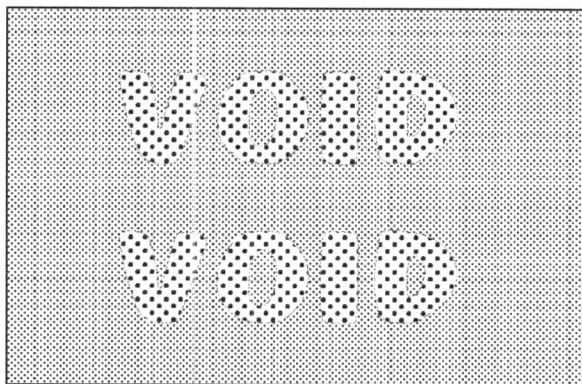

**Figure 5.2 -** A 'copy void' screen built into a design to detect copy attacks- the void message is invisible in normal view

For effect then, it is possible to produce what is termed a 'void pantograph screen' by creating two thicknesses of lines or sizes of dot within the image to be protected. The thinner lines or smaller dots are set below the finest level of scanner resolution and the larger dots just above this. Since some loss of quality through dot gain or drop out is inevitable during the scanning process, protected images will display the word or pattern chosen to alert that an illicit scan has taken place.

Since scanners, copiers and computer desk top printers are under continual development the process of embedding a void pantograph image into a label or carton design should undergo testing, evaluation and changes in color to obtain the best results. Since any scanner or copier can only 'average out' images that it can identify, the designer can compensate for this weakness by using subtle pastel shades balanced by stronger denser color at opposing ends of the spectrum to ensure maximum effect from such copy protection screens.

Finally it should always be recognized that depending upon only a single anti-copy feature in any design is in itself a weakness and other complementing copy detection features such as metallic inks provide a good balance of security for little extra investment.

## REFINING A LINE SCREEN TO PROVIDE HIGHER LEVELS OF ROBUSTNESS IN ANTI-SCAN SYSTEMS

A further more refined approach to detecting copies of labels and indeed any other form of packaging that carries a printed image is known as hidden image technology (HIT).

HIT involves the use of selectively pitched lines that carry a moiré interference effect and deliver a hidden message when print containing the covert embedded HIT image is viewed through a clear filmic filter. That filter carries a matching series of lines pitched in such a way that they interact with the printed screen to deliver the message.

**Figure 5.3 -** The image (left) shows an embedded covert HIT design- slightly enhanced so that it can be seen in the illustration. When a specially created viewer – which carries a correspondingly reactive screen, is placed over the print, a confirmation of originality can be observed

This system provides an easy to use tool that is both low cost and requires no training for teams involved in brand protection inspection activities. HIT works through the fineness and quality of the print screens involved in the original print operation and is robust against scanning attacks as any attempt to copy a protected image will involve a certain degree of dot gain in the copy. This superficial dot gain blocks out the image in the viewer, revealing the fake copy.

Further refinement to the system involves a scrambling of the original hidden image (or indicia) in such a way that it can only be viewed through a clear prismatic lens. Both lenses and origination software can be combined to provide an individual system for each user or brand owner. When combined with other overt and covert technology such as a hologram this is seen as the ideal defense against copy and replication attacks.

Alternatives to conventional HIT printing involve the use of polarization filters and specially manufactured foils or filmic materials that carry no visible image but reveal a security message or logo in negative or positive form when viewed through a special filter that acts in a similar manner to polarized sunglasses that remove the glare from the surface of a lake allowing the viewer to observe the fish swimming beneath the surface. This process is tightly controlled and 100% effective against copier and scanner attacks.

## ADDITIONAL REFINEMENTS THAT DETER COPIER AND SCANNING THREATS

As has been disclosed in the preceding paragraphs, something that interferes with the fixed resolution of a scanner or color photocopier can be used to protect printed images from such copy assaults. Since some printing technologies can be used to reproduce very small, almost nano-sized images they are ideal for the production of micro-text down to sizes lower than 1/100 of an inch. At this level of microscopic printing a 100 page text book could be fitted on to a sheet of paper 8.5 by 11.0 inches.

Lithography, gravure and intaglio are particularly suitable printing processes for creating micro-text images. Whilst intaglio is not a print procedure that is familiar in the label industry, some suppliers can be found that offer this highly secure process for brand protection applications such as swing tickets, tags and certificates of authenticity.

Litho and gravure are of course recognized as purveyors of high quality images and by using specific fine line micro-text typefaces and spot colors the process renders scanners and copiers useless for reproducing accurate facsimiles.

Finally, the ability to add some form of tactility to an image will result in the negation of copy attacks since they are unable to reproduce the physical properties associated with perspective and depth. Tactility can be introduced to printed paper and board through embossing or by adding a line-work screen from a finely etched plate that transfers a raised ink image to the material.

Such images are termed 'latent' (as in hidden) but are observable without any need for a viewer as required in HIT systems. Embossing or etching is used to create a series of micro-grooves that reveal an image when a label or pack is tilted towards or away from the observer. Light interference patterns created in the artwork result in a clearly observable image that can be 'flipped' between positive and negative states when tilted away from or towards a light source.

In order to visualize the effect provided by a latent image it is necessary to envisage a freshly plowed field where the furrows are invisible from high above but become observable at eye level when their peaks and troughs can be seen in high detail.

The effect of a latent image can be seen on the illustration (Figure 5.4) where a combination of fine line embossing and litho-printing has been used to create a leaf symbol that can be tilted to reveal a more complex and markedly different image when observed from different angles.

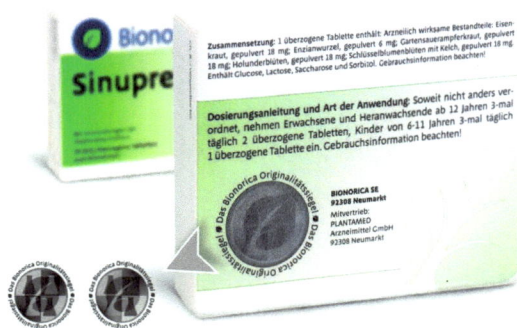

**Figure 5.4 -** Latent images built into the artwork provide a visible check on tilting to light. Such devices are robust enough to detect copy and scanner attacks

Such latent images are popular with high security print applications such as banknotes and identity documents such as passports.

These are produced using the intaglio (raised) printing process. The introduction of latent images using the litho process is seen as a very positive development for brand protection applications that rely on labeling and packaging for their success.

## DESIGNING TO DETER AND DETECT REPLICATION (RE-ORIGINATION) ATTACKS

With the widespread availability of origination software in all its forms there is also the risk of replication attacks on vulnerable product packaging.

Such occurrences are seen as more serious than copier and scanner attacks since they utilize conventional printing and converting machinery to create fake labels and cartons. This process can deliver more convincing counterfeits since close to original materials are used such as inks, self-adhesive label stock, carton board and the like.

There is no shortage of capacity in the printing industry which can be operated clandestinely by rogue staff during non-working hours in order to produce fake labels or containers. The industry is also defenseless against second hand printing equipment being traded on the open market. Such printing and converting equipment is then used by criminals who employ retired or redundant print staff to manufacture fake labels and packaging on their behalf.

It is not unusual for legitimate converters to be approached by criminal gangs who pose as re-sellers or agents and purchase 'on-behalf' of brand owners. Such an underhand approach removes any skill necessary to produce the counterfeit materials themselves. In some parts of the world and in the USA in particular it is an offence to be caught in possession of fake labels and packaging. That's how seriously the problem of product counterfeiting is being taken by national governments and authorities.

Adding security to the designing process requires special software or at least the inclusion of add-on 'security' modules to Adobe or other artwork origination systems. These resources provide the software tools necessary to create a whole range of security features within the artwork that are not available to the commercial printing sector. Such software is only provided to bona-fide and trusted

| Protecting against replication | | | | | | | |
|---|---|---|---|---|---|---|---|
| Guilloches | Special rasters | Crystal patterns | Relief images | Microtext | Split duct | Duplex printing | Vignette |

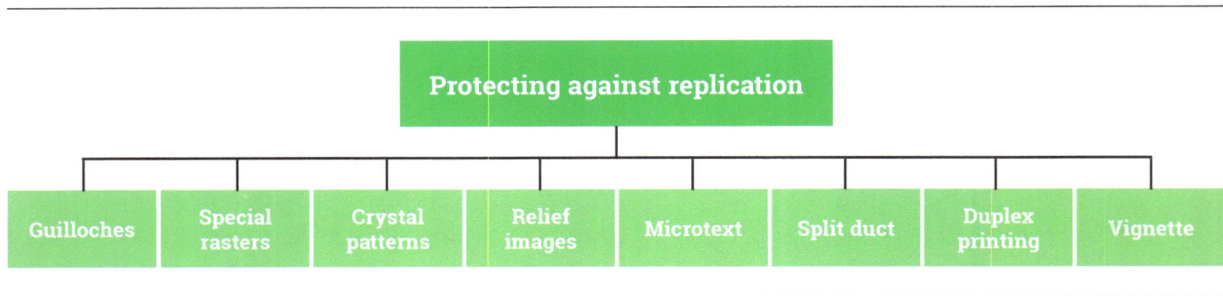

converters in order to negate the risks of such design tools falling into the wrong hands.

There are a number of intricate steps that can be taken to secure a design against replication and we will review a few of the most practical for brand protection in the next few paragraphs.

### GUILLOCHES

These features are most widely used in banknote design. They consist of geometric patterns of closely packed line-work that involves a number of colors for maximum security although single color printing of a guilloche together with an overprint of a spot color will provide a basic level of security on the simplest of label designs.

**Figure 5.5 -** Generic security design patterns used for protection against fraud or counterfeiting of printed products. Images courtesy of Agfa Graphics

### VARIABLE LINE THICKNESSES AND LINE MODULATION

This feature allows the designer to change the thickness of intersecting lines within a close screen in order to create an image within the screen that becomes evident when viewed from a distance in much the same way as a half tone made up from dots.

### SPECIAL RASTERS

Pictures found on the Web and photos you import from your digital camera are raster graphics. They are made up of a grid of pixels, commonly referred to as a bitmap. Security designers can design their own individual pixel shapes and then in conjunction with software tools use these pixels to create more complex shapes that can be 'grown' like a snowflake into interlinking backgrounds that change in hue as they progress across the design. Without special software these features are difficult to reproduce.

### CRYSTAL PATTERNS

This is the creation of complex symmetrical patterns that interlink and form a pleasing aesthetic effect and background upon which other security features can be built.

### RELIEF IMAGES

These effects are used to create the illusion that text is embossed or in a 3D font thereby adding a level of security through manipulating line screens into a series of curves and intersections that create the shadows and impression of relief printing.

## INTENTIONAL DEFECT

One popular technique that is designed to quickly identify a fake copy that has been re-originated is to introduce an intentional defect into the artwork. This may be as covert as a missing serif on a character in one line of text or a broken or damaged line in a frame around an illustration.

## MICRO LOGO SCREENS & MICROTEXT

The introduction of micro-logo's into a screen of micro-text and the ability to rotate and morph these into a background screen or as use as shading around a central object is also a method of making it very difficult for anyone trying to copy an original design through the use of all-purpose desk top publishing type software packages.

## SPLIT DUCT PRINTING

Popular in high security printing applications this skill involves the splitting of a central piece of artwork within the copy into two or more colors, but from a common plate. To achieve this objective it is necessary to add duct dividers to the ink troughs and rollers and add an offset roller to one color unit on the press. A different color is added to each split off section and when in operation the colors merge (where they intersect) into a pleasing mixture of shades that are unbroken in their hue and impossible to recreate without access to the necessary skill and resource of the original artwork and press used to produce them.

## PLATEMAKING/C'THRU DUPLEX PRINTING

Another skill that derives from the security printing preserve is the ability of a plate-maker and printer to co-ordinate impressions on both sides of the sheet in perfect register (perfecting). This requires the capability of in-line duplex printing and the creation of a design or pattern (in two segments – one negative, one positive on alternate sides) that create a complete image that can be viewed when held to light.

## VIGNETTES

An alternative to split duct printing that is less secure but offers the designer with the ability to fade a

screen from a solid 100% coverage of ink progressively and smoothly through to lower levels of saturation. Also known as gradient or blended screens these features are difficult to scan or copy effectively so that they retain the smooth properties of the original.

**Figure 5.6 -** The 20 denomination top left hand corner of illustration is made up of two interlocking sections of artwork printed on alternate sides of the sheet (see slight differences in shade). This technique is particularly suitable for wet glue labels that are to be affixed on glass jars, clear containers or PPE bottles

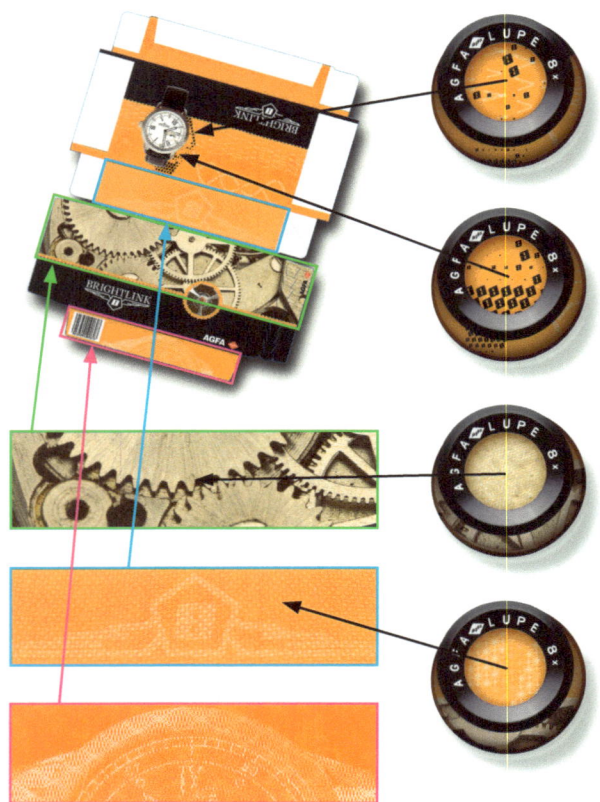

**Figure 5.7 -** The illustration shows various security designer skills and how they may be deployed in the protection of a carton board box against replication attacks. Source: Agfa Graphics

Please note that the printing technology used to produce this book is unable to reproduce all the security illustrations in the kind of detail that we would like.

Inevitably some quality is lost during image processing and our references to solid lines and solid spot colors may not always transfer to the illustrations accurately

## DESIGNING TO APPLY COVERT AND FORENSIC SECURITY FEATURES FOR TRAINED PRODUCT INSPECTION (DIGITAL WATERMARKS)

With the correct software installed it is possible to store a completely covert mark within any piece of artwork that carries a screen. This mark is secure inasmuch as it can be designed so that it is not possible to transfer the mark from an original to a scanned image. Such marks also protect against replication attacks where re-origination is involved, because any replica images will not carry the covert mark.

Such marks are termed 'digital watermarks' and should not be confused with their security paper counterparts. Neither should they be confused with the watermarking software used to embed background images such as 'confidential' in word processing documents.

For security and brand protection applications digital watermarking relies on a process that involves steganography. It is a technique designed to secure a message by hiding that message within another object so that it can be kept secret from everyone except the intended recipient. This is quite different from cryptography that renders the message (which is typically visible or audible) unintelligible to unauthorized viewers to prevent access.

Digital watermarking can be applied to both analogue and digitally printed images, although it delivers different attributes in each procedure.

Digital watermarking is achieved by varying the size and shape of some of the pixels that are used to compile the printed image. This is achieved by running the pre-press artwork file through a watermarking generator which applies the covert image chosen and invisibly embeds this in the pre-press artwork. When printing occurs the covert mark is replicated in every image produced from the protected printing plate.

The digital watermarks can be extracted from the protected image through the use of a scanner, or the camera on a smartphone equipped with a suitable app.

Both scanner and smartphone need access to the

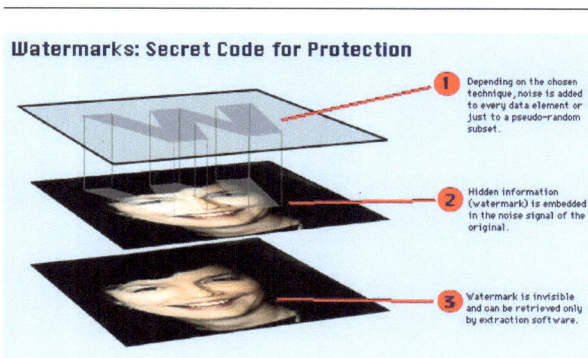

**Figure 5.8 -** Adding a digital watermark to print

**DESIGN ENCOMPASSES BOTH 'SYSTEM' AND CONSUMER ENGAGEMENT TOO AND REQUIRES MAXIMUM ATTENTION TO THE 'REAL ESTATE' (SPACE) AVAILABLE ON PACKS AND LABELS (ON-LINE PROTECTION)**

Chapter one identified and deliberated on the major tasks of packaging. These were those of protection, containment and as a carrier of information.

Labels too are required to carry information and as has been revealed, both of these components can be exploited to carry evidence of authenticity; evidence that can be tested in real time and if needs be forensically.

The primary task of packaging and labeling is to inform. In most cases where counterfeiting or diversion are considered major threats to a product line, this requirement carries the responsibilities of communicating the brand effectively together with a record of nominal data such as sell-by date, use-by date, identification (universal product code/lot number and source of origin) etc.

Additionally, in some market sectors such as pharmaceuticals there is a requirement to carry a unique serial number in order to identify each product and thereby create a pedigree that allows for provenance to be tested at every stage in the supply chain and again at point of purchase/use.

In applications like these, the informed designer will have to ensure that other important information such as a listing of the contents and directions on product safety and usage are not impacted through space restrictions triggered by the requirement for overt authentication and tracking devices.

Therefore, data carrying systems that require little or even no space such as digital watermarking, embedded NFC tags and the use of covert inks such as UV for the embedding of tracking data provide a useful compromise to the pressures of real estate on packs and labels.

In future it may well become necessary for each and every product to carry its own unique identity. If the benefit of the Internet of Things (IoT) is to be realized then such technology may well become mandatory from a practical point of view so that each and every item may be identified securely and tracked during its journey to final use.

appropriate extraction software that is linked to the original embedding process. Extraction of the authenticated image can take place visually in which case the covert image is converted to an overt form that can be viewed normally, or it may be programed to deliver a message or audio response of recognition to a screen on a laptop, mobile phone or tablet.

Whilst digital watermarks applied to analogue print will always display the same image on every copy reproduced, with digital printing it is possible to embed a watermark that changes with every copy generated. Therefore, digital watermarking in digital printing applications is an extremely powerful tool; since it can be made to deliver, say, a covert serial number in every piece of print created.

One distinctive benefit of digital watermarking is that it can be utilized to embed valuable supply chain information within the printed artwork on a label or carton. Since space is always at a premium on product packaging such as consumer goods, the availability of a process that allows additional information to be carried on products in this way is seen as a real benefit, since it offers the dual advantage of authentication linked with factual supply chain data.

# Chapter 6

## Applications for print security

Since it is necessary for all branded goods to carry a predominant version of the product name, and in many cases the manufacturer's masthead in the form of a prominent logo in corporate colors, the need to decorate and illuminate the label or pack requires the services of a printer and/or print converter.

The inclusion of the separate converting process in addition to printing will become evident in proceeding chapters and relates to the additional processes required to add value and functionality beyond the application of ink to material used to protect, identify, seal and carry the product between its manufacture and final consumption or use.

Whatever quantity of a product is produced, it will be necessary in the majority of cases to mark the merchandise in a way that provides recognition of its manufacturer, a list of ingredients or content, the volume, quantity or weight, health or safety requirements, nutritional information and provide suggestions as to how best to use, consume or handle the goods. Most consumer-related labels and packaging will also be required to carry a barcode – or even barcodes – and to meet all the relevant labeling regulations that often differ from country to country.

Obviously the requirement for small quantities of labels and packaging will require different treatment to those that call for millions or even billions of containers or decals. Furthermore, the constraints provided by the physical characteristics of the product will have an effect on the containers used.

Liquids will require bottles in PET or glass;

powders or many food and DIY goods will demand flexible packaging or cartons and compacted goods and components will call for direct labeling as well as secondary packaging materials. In many cases the product may be in a bottle or tray, inside a carton or carton sleeve, and also contain a leaflet or usage instructions in the pack – all needing to match with the same bar code or security-related information.

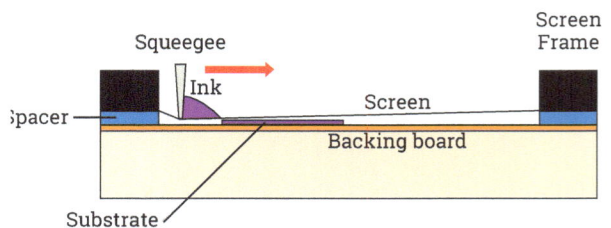

**Figure 6.1 -** The quality of printing on these infant formula containers enabled inspectors to tell the difference between fake and genuine product

In the pharmaceutical market there will be a requirement for blister packs, while for the protection and display of hardware, tools, batteries, cabling, etc., it will be essential to supply such items in clam shells.

These wide variations in labeling and packaging requirements from industry to industry are by no means meant to be exhaustive. They are provided as an indication of the wide envelope of services and solutions now available to brand owners world-wide. As packaging and labeling become more intelligent in future, this complexity is only liable to broaden even further.

During the last couple of decades printing processes have advanced enormously. We now have many new processes such as digital electro-photographic printing using both liquid and dry toner, both solvent and water-based inkjet and lasers for direct product marking that supply our growing need for personalization, low print runs and systems that can mark three dimensional products using inkless technology.

At this point it may be apposite to take stock of the industry and the traditional print processes available, together with a brief look at how some of the latest technology can be integrated and drawn into a seamless process that can deliver labels and packaging that addresses all of the complex security requirements of the modern brand owner. It should also be mentioned that in some added-value printing sectors, such as self-adhesive label printing, that combinations of printing processes may be used in the same machine to create a production line that encompasses, say, screen process, offset litho, hot-foil printing and UV flexo varnish.

Combination process printing such as this may also help to deter counterfeiters because of the high cost of press investment and the skill required to operate such equipment and integrate the various operations necessary to meet brand owner demand.

## LITHOGRAPHY

Also known as offset printing, lithography is a process widely used for printing glue applied labels and also for the production of cartons and swing tags. Originally a sheet fed process it now encompasses reel fed (continuous web feed) for the production of labels on the roll and for in-mold labeling and also for printing metal sheets for later conversion into closures, boxes and lidded tin containers.

The printing plates are wrapped around a plate cylinder which delivers the inked impression to a rubber blanket roller and this is offset onto the surface of the material being printed, thus the reference to 'offset'. Fine lines and excellent screen quality conspire to provide a superior image and offset-litho machines can carry a number of colors which can be duplexed on suitable machinery if close face and reverse register is required.

Plate making is relatively simple. Thin sheets of light sensitive aluminum plate are exposed through an original film which is created from a computer imaging origination process (see previous chapter). The printed image is separated from the non-print area through the clever use of ink (oil based) and water. The image to be printed obtains ink from ink rollers, while the non-printing area attracts a water-based film (called a 'fountain solution'), keeping the non-printing areas ink-free.

### Advantages of offset litho printing (for security applications)

The process delivers a consistently high image quality which can carry both fine line-work (which is an important tool for deterring replication and copy attacks) and covert HIT images within the structure of background screens.

Many wine and spirit labels are printed using this method as are health and beauty aids and personal care labels and packs.

Offset printing produces sharp and clean images and type more easily than letterpress printing because the rubber (offset) blanket conforms to the texture of the printing surface. The production of printing plates is a relatively quick and easy process especially if computer to plate technology is installed.

Properly developed plates running in conjunction with optimized inks and fountain solution may exceed run lengths of a million impressions and since each impression will deliver a high number of individual label or carton pieces it is possible to produce high quantities of quality product relatively cheaply.

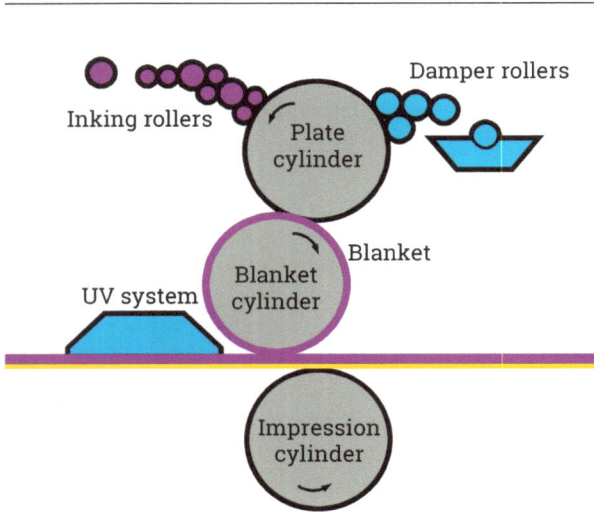

**Figure 6.2 -** Schematic of an offset litho printing press or print unit

### Disadvantages of offset printing (for security applications) when compared to other printing methods include:

Because only a thin film of ink is applied during the process there is a slightly inferior image quality compared to that of gravure printing.

There is said to be a propensity for anodized aluminum printing plates to become sensitive (due to chemical oxidation) and print in non-image/background areas when developed plates are not cared for properly.

The time and cost associated with producing plates and printing press setup makes smaller quantity printing jobs largely impractical. As a result, smaller printing jobs are now moving to digital offset machines.

### LETTERPRESS

Modern letterpress printing uses photo-sensitive polymer plates on which a raised image (ink) carrying area is carried. Plates are produced by photographic and direct image platemaking techniques.

Polymer plates are flexible and can be used in flatbed short run machinery (sheets) or longer run web

fed rotary presses where each cylinder print unit applies a different color to the web material. It is also necessary to dry the ink on faster running machinery with the assistance of ultra-violet light curing systems to prevent ink set-off and to ensure each color is cured before the next is laid down.

Letterpress printing is still relatively common in older roll-label presses, but with UV flexo mainly superseding its letterpress counterpart in recent years. Presses tend to be narrow-web, that is to say 200mm (7.5"), 250mm (10"), 360mm (14"), 400 mm (16") or 450mm (18").

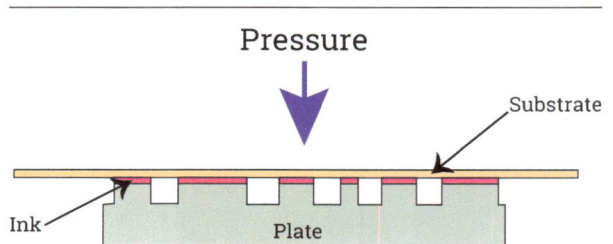

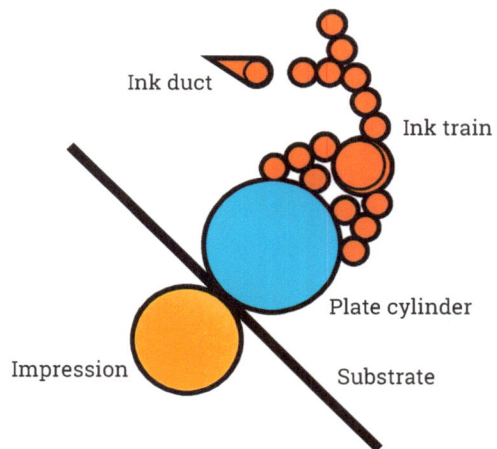

**Figure 6.3 -** Schematic of the letterpress 'raised' printing processes

The process was historically the most popular for printing labels and cartons but some would argue that the process is not as good at delivering quality illustrations in line or half tone format as its competitors.

There are two distinct versions of web-fed letterpress machinery in use in the label and packaging industry. These are termed semi-rotary intermittent and full rotary. Both processes are web fed but the former process feeds the web intermittently, briefly stopping to allow for die-cutting. Common impression and stack type letterpress machines are still used in certain markets and applications.

### Advantages of letterpress (for security applications)

The process still dominates the print industry in many parts of the emerging and developing world so is readily available to global brand owners that print locally.

In semi-rotary presses waste is reduced both in set-up and running the press since there is a straight though web-path which also allows for register to be maintained more easily. It is also cheaper to produce cutting tools for semi-rotary presses than full rotary machines. Cutting can be carried out online as can the application of metalized foil which cannot be achieved with litho presses.

Some would argue that there is less skill required to run these presses when compared with other machinery such as litho and gravure.

### Disadvantages of letterpress (for security applications) when compared to other printing methods include:

Full rotary letterpress is achieved from standard sets of print cylinders. In order to offer a range of sizes it is necessary to have a wide stock of cylinders and tooling available. This is an expensive investment.

The ink carrying ability and coverage with letterpress is not as good as gravure, neither is the final quality. Dot gain can be a problem too as letterpress plates age and wear. Therefore printing covert security images that require both consistent and accurate dot formation and size is a constraint.

The process is also said to underperform where large areas of a single color are being applied so that uneven coverage results in areas of wash-out.

Finally letterpress has limitations if optically variable inks are a requirement. The process is not able to carry enough ink weight to enable the dynamic color changes required in brand protection applications.

## FLEXO AND UV FLEXO

There are a few differences between letterpress which uses a raised printing surface and flexo which also delivers print from a elevated surface. The main difference is in the plates, the ink, and in ink metering in order to allow reproduction to take place.

For letterpress printing, viscous drying oil based ink is used whilst the flexo process uses water-based or very fine solvent based ink. For flexographic printing a softer, more flexible plate is used and this meshes with an ink metering system that delivers ink via an engraved anilox roller that carries the ink in thousands of small engraved cells.

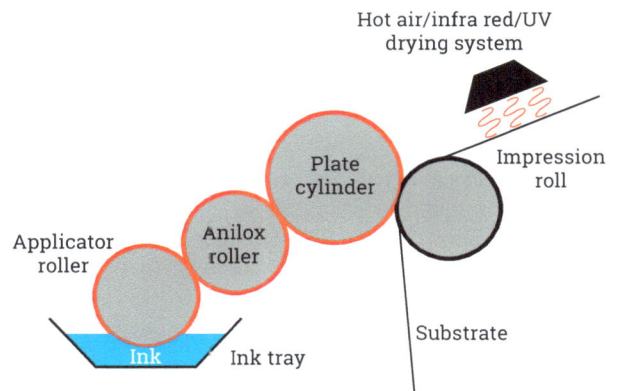

**Figure 6.4 -** Schematic of a flexo print unit. In UV flexo configuration a UV dryer station would follow this unit

Narrow web flexo is by far the most popular process worldwide for the production of self-adhesive labels. Some estimates put the percentage of installed flexo label presses in the USA at about 75-80% with Europe carrying around 55% of label capacity devoted to flexography.

The process is also popular for the production of wrap-around film labels, shrink sleeves and the printing of flexible packaging, sachets, pot lids, pouches and cartons.

Added to this, it is possible to apply scratch-off panels in order to provide some degree of customer interaction with the label or pack surface. (Such panels when removed by scratching reveal on-pack promotional games, authenticity checks and other consumer interactivity benefits).

The flexo process has been further refined with the introduction of thicker inks that carry more weight (opacity) through deeper cell volumes in the anilox roller and the ability to cure the ink using UV and LED drying systems. Lower dot gain is a benefit delivered by this system as is higher gloss and improved contrast on printed images.

### Advantages of flexo and UV flexo (for security applications)

From the plate making to the mounting process, to the transference of the image, flexography is as its name suggests: a versatile and adaptable means of handling large scale reproduction of images and text.

Flexo printing methods involve quick drying in a wide variety of ink types. Depending on the application and surface to be printed upon, users have the choice of five different kinds of ink. Solvent based solutions are ideal for sleeves, flexibles and other commercial uses, while water-based inks work well for more porous materials like carton board. However self-adhesive constructions and paper/board tend to dominate in this printing technology, where UV flexo is now largely the process of choice.

The process is ideal for printing clear and matt varnishes and especially varnishes that carry a pale optically variable or pearlescent pigment as an overt security device.

Quick evaporation of inks in flexography makes it a safe alternative for flexible food packaging.

Indeed it is important to note that any printer supplying to the food, drink and pharmaceutical sectors of the market is required to employ high standards of hygiene. Not only must staff follow these standards but equipment and the print floor must also comply with strict hygiene standards to ensure that pharmaceuticals and other items for human consumption are not contaminated with harmful substances. This requirement applies to raw material suppliers too. Inks that comply with food or drug contact regulations may also be required.

### Disadvantages of flexo and UV flexo (for security applications)

For long runs of flexible packaging materials gravure printing is considered more economical and is said to offer a higher quality than flexographic alternatives.

Some also say that flexographic printing cannot offer such complexity in artwork and design construction as gravure printing and that the density of color is also not as good. The development of High Density (HD) flexo has gone a long way to overcoming these challenges.

The set up costs for flexography get more expensive the higher the number of colors being printed and long web paths mean some material waste during make ready when compared with digital printing.

### SCREEN PRINTING

The image carrier for screen printing is a nylon or metal mesh; either in the form of a cylinder for rotary printing or a flat screen to link-in with say an intermittent letterpress machine. Indeed screen printing is more evident in the labels and packaging market as a combination process system than as a standalone in its own right.

Printing is delivered through photographically exposing the mesh to an image of the label required (color separated of course) and this either leaves holes in the mesh where ink can pass or blocks holes to act as a barrier to ink transfer. In this way the process delivers a 'screen' of ink to the substrate.

Today the process is used to print labels (commonly in combination with offset) for high added-value cosmetics and toiletries packaging, as well as pharmaceutical applications such as blister packs. It is possible using this process to achieve a highly controllable coating of ink to the substrate allowing for large areas of color to be applied without 'wash-out'. Ultra-thick ink films may also be used to achieve raised effects and/or enable Braille printing.

The process is highly suitable for the application of high quality optically variable inks since the ink weight applied can be varied in order to deliver very

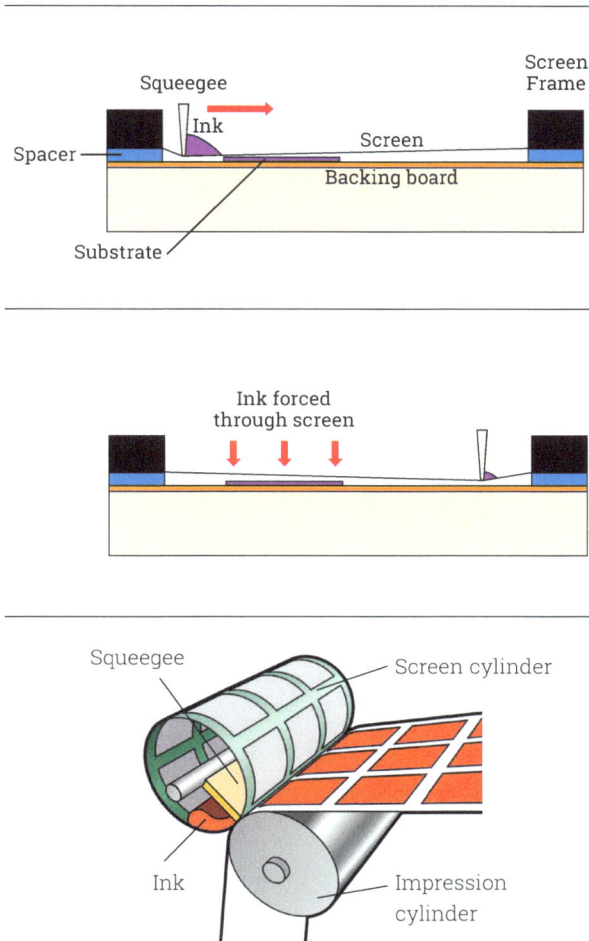

**Figure 6.5** - The illustration shows flat screen printing and rotary screen printing

pronounced color shifting effects. It can also print a high quality white opaque image, a process other label printing techniques find difficult to deliver.

### Advantages of screen printing (for security applications)
The major advantage of screen printing is its ability to produce large areas of perfect color and if spot colors are used or specific pantone shades are required, it can be difficult for counterfeiters to copy these exactly using other processes. The ability of the screen

process to deliver truly optically variable image (OVI's) effects is also a great strength. It should also be noted that screen printing is able to place a light color over a darker color without discernable loss of quality.

Also the process is flexible enough to be integrated into other print processes (combination presses) as part of a hybrid machine that can deliver a high number of individual colors, say CMYK plus a spot color and an optically variable effect (OVI's).

The process is ideal for converting stock metalized materials that are used to create cost effective holographic effects as part of the label or carton construction. In these applications a metalized material can be printed with a number of screen colors without any show through. This delivers an eye catching label that carries (a relatively low level) of overt security.

Generally screen inks have a high resistance to fading and can be used with confidence in applications where sensitivity to UV light or where rough handling are important factors. (i.e. labels for use with chemicals, insecticides, paints and other outdoor applications such as power tools).

### Disadvantages of screen printing (for security applications)
Flat screen printing is relatively slow and the process as a whole uses a great deal of ink. In security applications where OVI's are being used to create color change effects the costs can be high if large areas of ink are used for this purpose.

The process is not really suitable or cost effective where quantities over a few hundred thousand are required – unless the label or carton is particularly small.

Screen printing is relatively low in resolution so cannot be used for fine lines, masking hidden images or micro-printing. If these features are required then it is best to consider using a combination machine to print these features first and then overprint with screen in selected areas as a final operation.

### GRAVURE PRINTING
This is by and large a simpler printing process than flexo or litho as printing takes place directly from the inked image carrying area (which is cylindrical) and no

offset rollers are required to carry the ink to the substrate.

Gravure is not widely used in the narrow web label industry (other than in some coating applications) since it requires low viscosity inks that carry a high solvent content in order to dry them quickly. The solvent is highly flammable and requires a complex and expensive removal process in order to remove the risks of fire and explosion.

The process requires engraved print cylinders that absorb the ink in small cells and cavities that are reproduced used a photo etching process. Today, these cylinders are covered with print sleeves but previously heavily engraved print cylinders where used which were directly engraved and therefore very costly.

a prime benefit when such attributes are called for because foil is expensive to apply to small areas and wastage is expensive factor to consider. Metallic inks are also an effective deterrent against copier and scanner attacks since they cannot be replicated photographically.

Because of the ability of the process to deliver controllable depths of ink, the method is also useful to carry thermochromic inks that change color with temperature as well as scented inks and even holographic/pearlescent effect inks that display a prismatic light scattering image when viewed at different angles.

Finally gravure offers photographic quality graphics and the ability to transfer fine detail and skin tones to the substrate. This is because high resolutions are high (up to 1000 LPI) and dot gain is low.

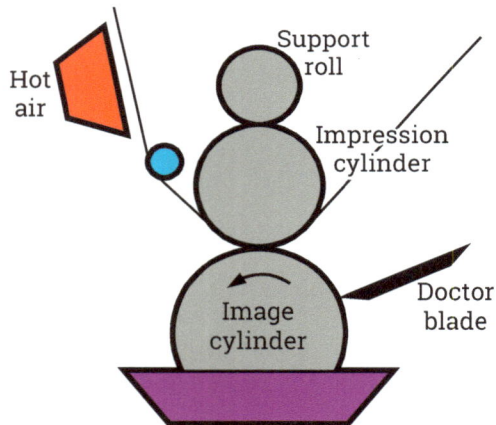

**Figure 6.6 -** Schematic of a gravure print unit

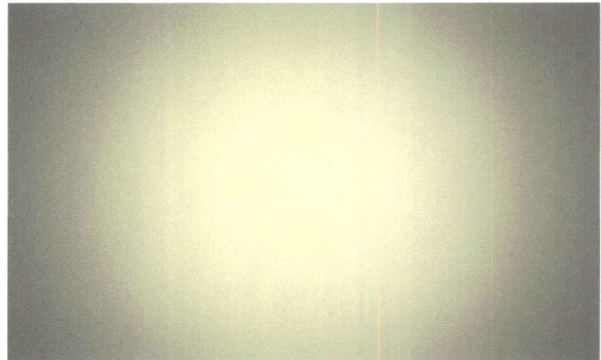

**Figure 6.7 -** The process is particularly suitable for the inclusion of security vignettes

Consequently the gravure process is more responsive to wide web long run work such as that found in flexible packaging, shrink sleeves and very large quantities of glue applied labels for consumer goods such as those found in the beauty and personal care markets, beer and beverages too. In these applications, where high numbers of colors are required above the basic CMYK gamut gravure excels, especially if metallic inks are specified. With gravure printing metallic inks carry a high shine and are often difficult to distinguish from metal foil. This is

### Advantages of gravure (for security applications)

In security related printing applications, gravure is seen as a process of choice for long runs where there is little likelihood of copy change and the cylinders can be re-used again and again. The ability of the process to carry high resolution images is also an important factor as well as the capability to deliver photochromic and optically variable inks and coatings.

The process is also highly suitable for the transfer of scratch off inks for security applications where covert coding is required. Highly secure scratch off coatings can be achieved using a layer by layer approach where two or more cylinders are used to build up a coating of scratch-off latex. This is why gravure is the process of choice in security applications such as instant lottery tickets.

It is possible to add gravure stations to existing flexo presses if there is a demand for mirror like metallic or other features such as photochromic reactive areas on a label or pack.

### Disadvantages of gravure (for security applications)

Where highly solvent inks are required the expense of making the print hall and environment safe from the risk of explosions is a deterrent when compared to other much safer processes. There are also environmental considerations too.

With gravure there is also a costly upfront investment in sleeves which are more expensive to engrave than polymer or litho plates. Most converters find it more cost effective to sub-contract the engraving process and thereby lose flexibility and control in situations where a quick turn-round (or security) is called for.

With many label and carton applications calling for smaller and smaller quantities as brand owners prefer to change designs more frequently and apply just in time stockholdings, gravure is at a disadvantage because it skews in favor of long runs and no-change repeats.

### INTAGLIO PRINTING

Associated mainly with high security printing applications such as banknotes, some high value postage stamps (where they are still used), excise stamps and passports, intaglio printing is rarely used outside these fields. This is because the process is highly controlled by the central banks that print and issue currency at state level and a few independent, long established high security print houses that produce currency on behalf of states that do not have their own dedicated banknote production capabilities.

Intaglio produces a raised tactile print surface from an engraved cylinder like gravure. Indeed both processes are very similar as they deliver very high print quality with low dot gain and when the tactility is

added to the process, an overt security feature that involves 'feeling' the raised print is provided.

**Figure 6.8 -** Intaglio print produces a tactile surface and high resolution print quality and is suitable for combination with other processes such as lithography. NOTE the split duct effect of orange/purple on opposite sides of the label. These color variations were delivered from the same printing plate

The process tends to use deeper engraving and more viscous inks than gravure in order to deliver this raised effect and to enhance sharpness and tactility further the image is forced into the substrate with a great amount of pressure. Both a doctor blade and a wiping paper are used to remove surplus ink from the engraved print cylinder in order to deliver exact amounts of ink cleanly to the substrate.

It is usual for printers of banknotes to overlay the intaglio print on a foundation of litho background screens. It is also possible to print an intaglio image over the top of foiled security holograms providing a very secure combined security device.

It is conceivable to produce pressure sensitive labels using intaglio but the pressures involved in releasing the ink from such deep recesses in the plate can cause problems with unwanted adhesive egress. That said there are a few label suppliers in North America and security printers in Western Europe that can offer security labels and seals that take advantage of this process.

### Advantages of intaglio (for security applications)

Because of its ability to carry high levels of ink, intaglio is ideal for delivering security features such as optically variable inks and latent images. The depth of the ink combined with the shape of the fine line engraving can deliver an effect that allows a latent (hidden) image to be viewed when tilted to light.

Intaglio is also able to produce microscopic lines of type because of the pressures involved, as well as 'blind' embossing at the same pass. Intaglio is also capable of delivering excellent split-duct screens.

The main benefit of the process is its relative scarceness, which may be why it is attractive as a process for securing certificates of authenticity and a few very high security applications such as swing tickets for designer goods.

For high volume products such as passports, postage stamps and banknotes where designs may remain unchanged for many years the run-on (repeat) costs associated with platemaking are economic as the plates last for millions of cycles.

### Disadvantages of intaglio (for security applications)

The main drawback is that set up costs and origination/engraving require a high initial investment so that small, and even what would be large qualities for other print applications such as flexo, are uneconomic for intaglio.

Because of the high resolutions involved and the depth of engraving it can be a lengthy process creating plates. Since the call for intaglio in the product security sector of the market is so low equipment can be left idle for long periods between jobs.

The process is pretty well confined to a relatively small number of suppliers of machinery, and because of economies of scale the print machines and associated pre-press processes are expensive too.

### DIGITAL PRINTING

Over the past 20 or so years, full color digital printing has evolved to become a major contender in the label and packaging markets where the toner based technologies successfully compete with offset and combination presses in the higher-end markets, and with inkjet challenging flexography in material runs up to around 5,000 linear feet or so. However this restriction is often invalid when considering digital print for security related applications.

**Figure 6.9 -** Digital printing is suitable for producing variable codes and for personalizing labels and packaging. It can be combined with flexo, litho and letterpress to improve authentication systems and add track & trace mechanisms

This is because digital is the only print technology capable of adding secure serialization in visible and digital watermarked format, to labels and packaging. Granted that numbering boxes can be used to apply serial numbers that can be used to track and trace product in conventional print applications, the fact that this process carries no security and can be easily compromised adds to the attraction that digital offers; the ability to provide personalization, customer interaction and security in the same combined process (see Figure 6.9).

There are almost as many print technologies now available in digital format as there are conventional print techniques. Ink based digital offset (HP), toner based systems (Xeikon – now owned by Flint Group), inkjet printing and heat fusion all conspire to offer attractive options for brand protection applications.

Today, we also find digital present in hybrid presses so that the benefits of conventional and digital print can be combined into a process that embraces the attributes of each technology allowing for high resolution color and graphics to be displayed alongside customer centric functions such as interactivity and authentication. This also includes the ability for a brand owner to enjoy total asset visibility (track & trace) throughout the entire supply chain.

## Advantages of digital printing (for security applications)

The major advantage of digital is that no printing plates are required, so set up is easy with little wastage and the ease of entry it offers into what had previously been a relatively protected market that required high degrees of skill and dexterity to run a printing press.

This means that there is no requirement to invest in a variety of print cylinders because the image produced can be of variable length and no plate gaps are called for. However, there may be a requirement for some additional investment in enhanced pre-press, digital front-end, color management and workflow stages. Good color management is seen as crucial to the success of digital printing.

The introduction of extra colors (beyond CMYK) to the digital tool box means that the color gamut has been further enhanced to compete with conventional print. Add to this, the increases in resolution that have been achieved, now make the process almost comparable quality wise with other traditional forms of printing such as litho, flexography and letterpress.

Of course, the most important ability of digital has been previously mentioned. This is its capacity to provide variable print and graphics, sequential codes and numbers, batch and date codes, personalization, and deliver individual copies of a common design across a range of substrates including pressure sensitive, flexible substrates, metal, carton board and paper.

## Disadvantages of digital printing (for security applications)

The main disadvantage of digital is that it runs at lower speeds when compared to flexography and other methods of analogue printing, although running speeds are now significantly improving on the latest generations of digital presses up to 50, 60 or more meters per minute), However, the speed can be slowed in high resolution jobs by the requirement to print variable images or produce complex jobs where information is called from digital files and then placed within selected print areas on the substrate.

There has historically been a very high cost associated with the acquisition of digital print presses and also a high cost related to consumables. Furthermore, finishing operations such as waste stripping, die-cutting and foiling until recently needed to be carried out off-line. New developments in automated digital finishing and in laser die-cutting, however, are now enabling increasing numbers of digital presses, particularly the higher-speed inkjet machines, to run with in-line laser die-cutting and digital finishing solutions.

The cost of investing in inkjet is has also come down dramatically in the past few years, with some stand-alone 4-color process inkjet presses now available for purchase for as little as Euro 40,000 (US$ 45,000).

To overcome some disadvantages – such as a restriction on the types of security ink available (i.e. optically variable and photochromics) and its inability to print solid lines – digital in security applications is better confined to combination printing processes where the benefits of both processes combine their better attributes which enhance security further.

## A BRIEF LOOK AT 'NEAR FUTURE' PRINT PROCESSES AND THEIR POSSIBLE IMPACT ON SECURITY RELATED LABELS AND PACKAGING

One of the most promising upcoming print processes that is related to smart and security labeling and packaging applications is printed electronics. This encompasses the production of printed circuits and organic electronics that can be used as various sensors to alert consumers visually to some risks associated with product safety such as unauthorized opening, temperature sensing and authenticity checks.

Printed components can now include simple memory, sensors, logic, displays and the batteries to power these components.

As time progresses it is expected that such technology will find applications in pharmaceutical, food and drink and other packaging-related sectors where tampering and product status such as temperature and humidity control are important factors in product safety.

Whilst smart labels and security applications still only account for one percent of the total packaging

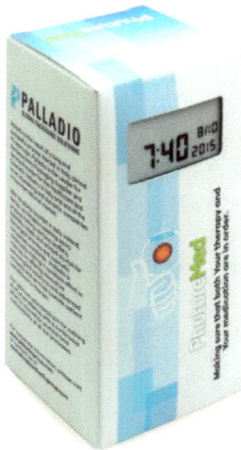

**Figure 6.10 -** The illustration shows the introduction of a variable (e.ink) display that provides important information on product provenance and safety. The display is powered by the red button (center) which is pressed in order to gain access to the information presented by the on-pack liquid ink display

**Important to note:**
Most of the advantages and disadvantages in the various printing systems listed above are subjective inasmuch as many press manufacturers will argue that their own processes are better than those of their competitors. What is clear however is that there are no known single print processes that are suitable to serve the needs and demands of all the brand protection applications out there, and across every sector of the marketplace.

and labeling markets they carry forecast growth rates of 15 per cent.

Developers and producers of such innovate products rightly expect better returns on capital invested than they can achieve from mainstream applications alone.

# Chapter 7

# Foiling, embossing and 2D, 3D holographic solutions

Most of the previous chapters have pointed out that qualities such as uniqueness, application of skilled technology and processes together with other barriers such as high investment in plant and equipment and proprietary processes, are all obstacles to counterfeiters who may wish to copy or replicate labels and packing in an attempt to fool consumers into purchasing fake products.

Other attributes such as the introduction of hidden anti-scan features and stealthy forensic markers are also representations of technology that may be used to identify bogus goods through their packaging and markings.

To be effective, and in order to act as a warning or precursor of the possibility that a product may be counterfeit it needs to carry easily recognizable, but secure characteristics, that can be identified as genuine by purchasers and those policing the distribution chain in order to root out fakes and diverted product.

In well balanced, secure counterfeit detection and deterrence systems, there will always be the presence of overt, covert and forensic technologies that are tasked with initial identification, secondary confirmation and finally forensic proof that may be used in legal challenges that require court action.

## FOILING AND EMBOSSING
Foiling and embossing are both supplementary operations that are provided by printers to enhance a products appearance through its packaging or wet glue labeling.

Though it is not usual to see embossing on pressure-sensitive labels because the base material is not conducive to high degrees of pressure, it can be delivered through intaglio printing with the right material present and even simulated by silk screen processes that mimic the relief and tactility that

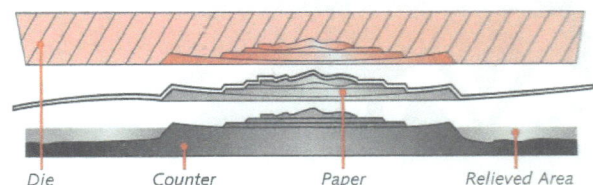

Die      Counter      Paper      Relieved Area

**Figure 7.1 -** The male/female dies - formation of an embossing process

makes embossing an attractive presence.

However, most embossing is carried out through the use of engraved dies (in male and female formats) that press (force) the paper from the front or back into shape. The process is more effective on carton board than it is on paper and produces a cleaner edge to each design or letter of type that is transferred.

Embossing is a popular form of application for mandatory braille that is used on pharmaceutical product packaging to assist sight impaired citizens in identifying their medication. In these instances adding a further embossed verification decal simultaneously is a no-cost benefit.

## TYPES OF EMBOSSING

There are different types of embossing process depending on the particular effect, image or design required and whether the embossing requires to be registered to prior letterpress, litho, flexo, screen or digital printing, or to hot or cold foiling. These different types of embossing process are summarized in Figure 7.2 and under the subsequent sub-headings:

**Blind embossing.** A blind emboss is embossing which has not been stamped over or registered with a printed image or with a foil. The color of the blind embossed image is the same as the color of the substrate surface. It can also be called a self-emboss or same color embossing. Quite simply, by creating a raised area using a die, blind embossing is able to create a subtle paper colored image that can be felt as well as seen, offering both visual and tactile appeal. It is especially effective when a subtly elegant, three dimensional image is desired. Different label materials, such as paper, film and foil, create different

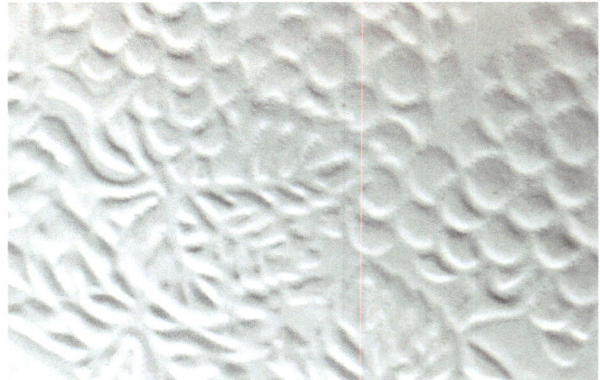

**Figure 7.3 -** An example of blind embossing

effects. So embossing can be a versatile option for creating a standout label.

**Debossing.** In this process, the substrate surface is depressed instead of raised as in conventional embossing. De-bossing uses the same techniques as embossing to create the necessary indentation, except that the process involves the application of pressure to the face side of the substrate, forcing the material downwards into the female die so as to create the recessed profile. This can be as emphatic or delicate as the graphics or words dictate, and careful choice of substrate (avoiding bright whites and very smooth materials) will undoubtedly enhance the effect.

**Registered embossing.** Registration of one or more printed colors across the embossed area enhances the process further and makes it even more

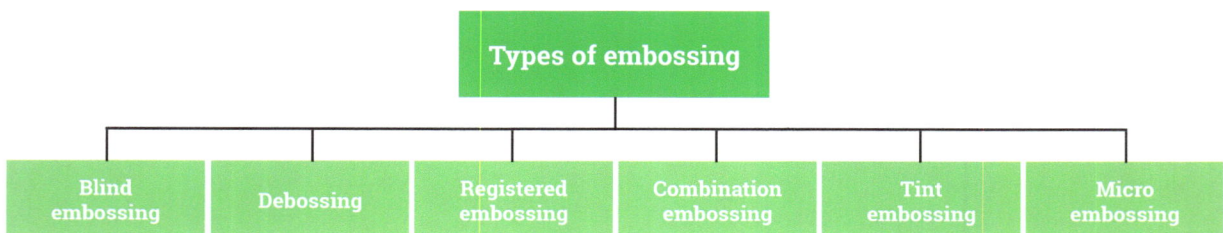

**Figure 7.2 -** Types of embossing used in the label and package decoration sectors

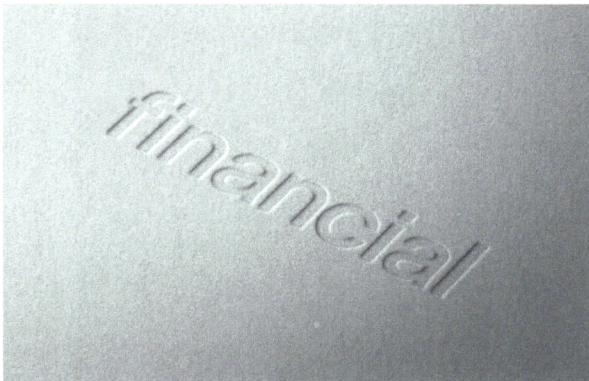

**Figure 7.4 -** A debossed image

**Figure 7.5 -** Registered embossing enhanced by an overprinting

scanning. Combining the embossing with the foil means that any replication attempts that use a traditional print approach require the skills and the tooling to enable the attack to succeed.

The process of foil stamping (Figure 7.6) can be carried out either off-line (usual in sheet fed applications) or on-line (reel fed processes). In on-line applications this requires a special foiling station to be installed on the press that works in combination with the other printing and die-cutting units.

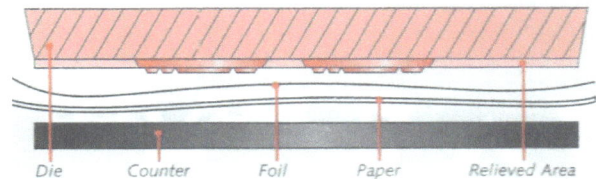

**Figure 7.6 -** Foil stamping provides the embossing process with the addition of a metallic finish

**Tint embossing.** A relatively new method of embossing in which pearl or pastel foil is used in the embossing process. The methodology is the same as other types of embossing but the technique has become increasingly in demand. For tint embossing it is generally best to use white stock because pearl and pastel foils are transparent.

**Micro embossing.** This is where the effect is achieved with minimal depth but using intricate and complex designs. The process has become increasingly attractive in security applications such as event ticketing, anti-counterfeiting and legal documents. Indeed, embossing may be used for a wide range of security purposes. Embossed seals or symbols of authenticity add security features to labels, government forms, legal documents, and corporate papers. Having said that, security embossing today is perhaps considered as an older form of document or label security, as more alternatives in analogue and digital print processes, origination and pre-press, foiling, etc., continue to be developed.

**Glazing.** This refers to a polished emboss.

difficult to copy, replicate or scan.

Therefore embossing and registered print are seen as a basic defense against counterfeit attack and can be widely observed in use on the labels and cartons of wines and spirits seen on the shelves in supermarkets, liquor stores, duty free shops at airports and in bars worldwide.

**Combination embossing.** Taken a step further, the addition of a hot or cold stamping foil in register with the embossed image is an effective method of deterring counterfeit attacks since the foil is impossible to replicate through copier attacks and

Glazing is a popular technique used on dark colored stock. The heat and the pressure when pressing the die are increased substantially. This adds shine to the surface. If a very high temperature is used, light color papers can be scorched to change the paper color. This provides for great contrasting designs if done properly.

As can be seen from the above summaries, there are many types of embossing to choose from depending on the embossing design created, the nature and type of substrate being used, the embossing effect required, whether the embossing is blind and raised or de-bossed and recessed, whether it is registered to prior printing or foiling.

A further embossing variation is die stamping, a method of printing and embossing the image using the same engraved steel or copper die. In this process, flatbed male and female dies are mounted in the same way as normal embossing dies and placed in a die-stamping press. This press usually has a letterpress roller inking system which deposits a film of ink onto the surface of the female die. The substrate to be die stamped is positioned between the two dies which are then pressed together under extreme pressure leaving a printed and embossed raised image.

## COMPOSITION OF FOILS USED IN PRODUCT PROTECTION APPLICATIONS

Most stamping foils are comprised of five layers. The top layer (furthest away from the print substrate) is a polyester based film carrier. Under this top layer is a thin film of release coat which allows the release of the foil from its carrier when it is impacted by the die. Next is a lacquer or color coat that carries a pigment that gives the foil its color. This layer is transparent or translucent and provides color which can be matched to a specific shade if required.

The forth layer is the metal coat which is generally formed from metalized aluminum and this provides the reflective qualities and opacity required to prevent see-through from the material below. The final layer is an adhesive coat that provides the bond between the foil and the substrate.

## HOLOGRAPHIC FOILS

Holograms have found a useful application in securing print against unwanted copying and counterfeiting activity since the early 1980's.

This is because holograms are a very useful overt recognition technology that can be adapted to each customer's individual requirements. The process has progressed over the years through simple 2D/3D designs which can be viewed as 'entry' level through to highly complex and counterfeit resistant designs that are continually evolving to combat the developing skills of the counterfeiter.

It needs to be recognised that security in print never stands still and is in continual competition with those who wish to compromise security devices and substrates.

Whilst there were only a handful of companies capable of producing holograms initially, this figure has grown to many hundreds today as universities and the growth of packaging and labeling in China and India drive the demand for holograms as a decorative feature as well as a security device.

Holographic products today can be separated into two distinct types, diffractive embossed optically variable image devices (DOVID's Figure 7.7) and photopolymer filmic based technology.

Photopolymer (reflection) holograms are manufactured from photographic type film and were indeed the first forms of holography developed. However, these early filmic devices were considered unsuitable for use in security printing applications because they were film based and needed to be coated with self-adhesives in order that they could be affixed to paper and board.

The process progressed further with the introduction of embossed holography which was more adaptable and could be applied in the same way as a traditional stamping foil. Since the printing industry was more familiar with this type of application technology and photopolymer holograms at that time were considered unsuitable for anything other than artistic purposes, DOVID's became the choice of many users who were faced with the challenge of protecting their labels and cartons from un-authorized coping and tampering.

Holography in embossed format relies upon the

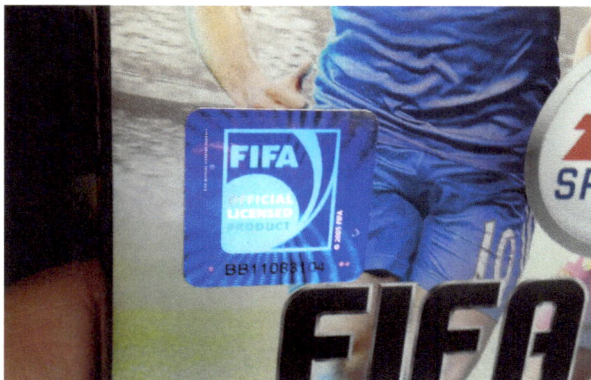

**Figure 7.7 -** An embossed hologram foil label used in product authentication

**Figure 7.8 -** Embossing the holographic foil

complex use of lasers and mirrors and lenses to record an image. Such images need to be originated just as conventional print requires origination too. Various origination techniques exist and some are more secure than others.

Origination of embossed holography utilizes a sophisticated computer-driven system to create images in an embossing die that can be stepped and repeated to fit a narrow (or wide) web embossing machine. Metalized foil (see above for foil structure) is passed through the embossing process (Figure 7.8 right) and the resultant finished multi-layered material is slit and delivered in rolls ready for application to labels and packaging using either off-line application processes or specialist in-line tooling (hot or cold foil blocking/transfer).

There are basically two types of holographic foil available to the designer, holographic patterned foil which offers a 'wallpaper' continuous repeating design (Figure 7.9) and bespoke registered holograms (Figure 7.11) which offer images that sit in isolation and must then be picked off the foil in registered with the print substrate.

Wallpaper or continuous type holograms are offered as stock items for separate foil application by the printer or already applied to paper and carton board for immediate conversion by those printers who do not already have a foiling capability. It should be

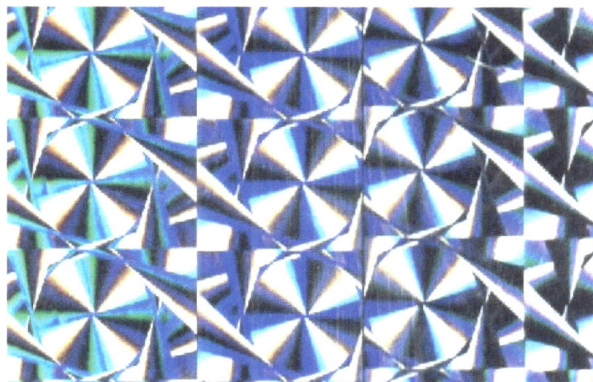

**Figure 7.9 -** 'Wallpaper' or continuous design holographic foil

noted that in reacy applied form, holographic materials are have a total surface area of holographic coverage and this again limits their use mainly to decorative applications.

These continuous holographic stock products are not really secure and are often seen in applications such as toothpaste packaging and other personal care items such as shampoo and mouthwash. The hologram's main function here is to act as an eye catching design feature on the shop shelf rather than a security device.

It should be noted though such holograms are also available in stock foil format and can be used

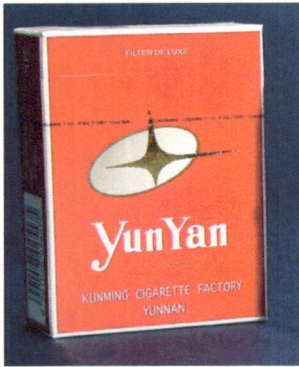

**Figure 7.10 -** Stock foil wallpaper hologram used in combination with a company logo/die

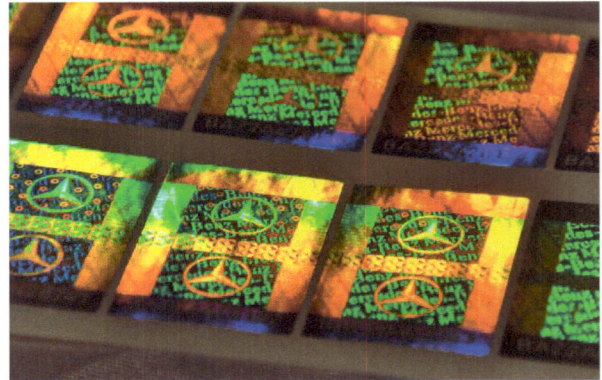

**Figure 7.11 -** Shows the bespoke production of holographic foil. Each logo and surrounding piece of foil needs to be 'picked out' during the foiling process this is achieved through the use of a registration process in combination with the foil blocking head

successfully used as a blocking foil in combination with a company logo die or as an embellishment to a blind embossing feature (Figure 7.10 right center of pack).

Bespoke registered holograms (Figure 7.11 Mercedes Benz illustration) are considered to be the most secure since they require the use of a registration mark in order that they are registered with the printing web and placed exactly in the same place on every piece of print produced.

This type of hologram requires careful design and accurate origination as it must be transported on the polyester carrier roll and placed exactly in position on a label or carton in order to provide conformity of product. Inexactly registered, or partially released holograms lead to uncertain authentication decisions later in the supply chain.

To prevent this from happening, each discrete hologram is produced with a corresponding registration mark aligned to its position on the carrier film. This registration mark is recognized by an optical read head placed on the foiling equipment and this automatically identifies the position of the hologram and activates the foiling head accordingly so the image is always placed in the correct position.

Such accuracy of production can only be obtained through the use of web inspection machinery similar to that used to control and identify print color

registration on the press. Cameras are used to inspect each piece of foil, compare this with a pre-enrolled quality image and reject non-conforming product.

Just as traditional security print can be protected from copy attacks by sophisticated design elements such as guilloche patterns and complex color registration, security holograms can be protected from unauthorized duplication by a range of in-built security design features.

- **CLR (Covert Laser Readable) Image –** This is an invisible embossed mark that can reflect an incoming narrow laser beam (as seen in laser pen pointing devices) and project a recognizable image from the CLR source onto another nearby surface. Such images may be simple designs such as a logo or OK symbol.
- **Concealed Images –** These are images that can only be revealed when the embossed hologram is tilted towards or away from the viewer.
- **Combi-Hologram –** As its name implies this is a combination of origination technologies

from different sources that combine to create one difficult to replicate image.

- **Computer –** synthetized 2D/3D – Embossed holography works through creating an image on a multi-planar surface similar to a reflector lens used in tail lights on automobiles. As the angle of view changes, light falling on the surface is diffracted giving the illusion of depth and movement.
- **Guilloche patterns –** In similar way to guilloche patterns used in traditional print origination these design devices create a complex series of intricate waves and radiating lines that change color and width when viewed at different angles.
- **Kinetic Images (sometimes called Kinegrams® and Excelgrams®) –** These images deliver parallax movement and bright dense color even when viewed at low levels of illumination. In more secure forms they are controlled and only seen on travel documents and banknotes. Tuned down versions that are suitable for brand protection are also available.
- **Microtexts and Nanotexts –** These act in the same manner as their analogue printing cousins. However microtext and nanotext in holographic form can change color and deliver kinetic movement from behind or from above another object such as a guilloche screen or logo.
- **Stereograms –** These are sophisticated, three dimensional holograms that combine a 'live' series of frames from a movie or moving model. Various views are shot and embedded in the origination providing an illusion of movement when the hologram is tilted or turned. Such images are impossible to replicate exactly since to achieve a satisfactory copy the counterfeiter needs access to the original movie or moving model.

Security holographic foils continue to develop in the face of constant challenge from counterfeiters. This is because holograms are one of the most popular, and secure forms of authentication and

provide all of the attributes necessary for accurate authentication (overt, covert, forensic). Therefore as their use widens in the fields of currency and ID protection and product security there will continue to be advances in new features and manufacturing processes in order to preserve the integrity of such devices.

More recent developments in this field include:

- **Ultra-high resolution imaging –** This kind of technology allows the creation of surface holograms with a resolution of up to 0.1 micrometers (254,000 dpi)
- **Introduction of 'pure' white and black to white color switch –** Up until recently it was not possible to deliver a 'pure' white to an embossed hologram. White light is a combination of all the prime colors and difficult to reconstruct using refraction gratings (which is the basis of a DOVID). Color switching from white through to black and all the corresponding gray scales between these two extremes provides a resilient, observable security feature.
- **Custom pixels –** Using ultra-high precision control, uniquely shaped pixels are micro-positioned to optimize the optical effect and provide an image that has a highly complex and recognizable forensic fingerprint.

Security can be further improved by the introduction of track and trace numbering which is carried out using laser ablation, thermal transfer or direct permanent ink jet. This process creates an image comprising a series of characters and/or letters making up a code that can be used to identify each unique hologram in the same way that banknotes each carry an individual number to identify each individual banknote (Figure 7.12).

The benefit of this process is that all production from each batch of holograms can be accounted for and audited to ensure there are no duplications or over runs in existence.

Further information on serial numbering and barcode application to embossed holograms appears in the next chapter.

One important point though: the more complex

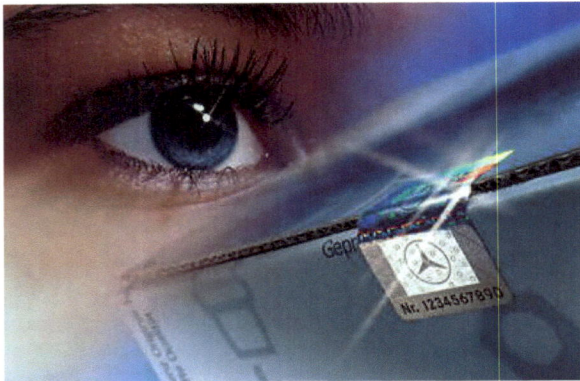

**Figure 7.12 -** Adding a serial number to each hologram during the embossing process provides extra auditing benefits and also the ability to track & trace the protected product

the hologram the more difficult it is (time required) to be authenticated if multi-level overt, covert and forensic layers are introduced. Using a few high level features at primary level so that quick visual confirmation is achieved is preferable to multifaceted hidden devices that require different tools to confirm their presence.

All that is generally required for brand protection applications is one secondary system such as an encrypted barcode or laser activating (CLR) type images. If it is necessary to move to forensic approval then a taggant type approach where a molecular marker is incorporated in the body of the hologram adhesive layer is a practical solution. This latter step is only considered necessary for items of extreme value or for financial and identification systems.

## TAMPER EVIDENCE AND DE-METALIZATION
There is a trend to incorporate holographic images into tamper evident labels, and this approach has merit when a carton or flexible bag needs to be sealed in a way that protects it from tampering and provides proof of first opening.

| 2D/3D | Secure 2D/3D | Computer originated dot matrix | Stereogram (live action) | Multi-gram (combination of processes) |
|---|---|---|---|---|
| Low level devices | Medium level security | Generated by special security computer software | Originated from 35mm film or video through extracting 'live action' stills | Combines any two or more recognized security processes |
| Little or no security | Originated from still models | Complex designs available | Can include computer generated components for added security | Dot matrix and stereograms |
| Widely available | Needs high levels of light transmission for verification | Can be 'proofed' on screen | Available from only from recognized & controlled security vendors | Can contain covert devices such as bar codes and IR messages |
| 1,000's of vendors worldwide | Seen on Credit & Debit cards ie Mastercard/Visa | For security applications 100's of vendors worldwide | | Restricted to high security use and only available from controlled and audited vendors |
| Used mainly for promotion Used also for design effect as continuous background on packaging | | Can be viewed in subdued light Design can include covert features | | |

**Figure 7.13 -** Various layers of security are available for holographic foils depending upon the value of the product protected

This is achieved by making the metalized holographic films frangible and introducing a tamper evident design into the adhesive layer (Figure 7.14) shows the effects of removal of a tamper evident holographic foil label).

**Figure 7.14 -** shows the effects of removal of a tamper evident holographic foil label

When a permanent adhesive is used any attempt to remove the label results in an immediate deformation of the hologram and attempts to reseal the label will be evident. If edge cuts are placed around the circumference of the label these too act as an effective indicator of any previous un-authorized opening.

Security for embossed holographic foiled images can be further increased by selectively de-metalizing the foil after the embossing process has been completed. This is a process much used in currency protection and applies a further skilled operation to the foil making it more secure.

De-metalized holograms can be seen on the UK £20 banknotes and on Euro banknotes above €50. The process allows for a much more intricate blocking configuration than that produced from a patterned blocking die and crisp images of micro-lettering and intricate border decorations can achieved that push the boundaries of successful copy attacks.

The effects of de-metalization can be seen on the image (Figure 7.15) from Holoptica. Note that the transfer of the image from the carrier web is completed without the need for a special patterned blocking die.

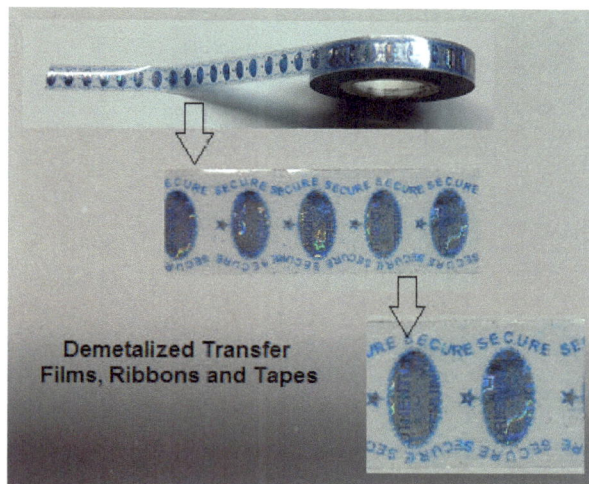

**Figure 7.15 -** Demetalizing the holographic foil to remove unwanted background material allows a very high resolution textural image to be placed on a label or pack. This process further complicates attempts to counterfeit the hologram

## SLEEVES, LIDDING FILMS, BLISTER PACKS AND TEAR TAPES

As a process, embossed holography is highly resourceful and has been applied to a variety of security tamper proof sealing technologies throughout the package goods market.

The technology can be seen in use on tear tapes for cigarette pack wrappers and in this form it provides a visual check of authenticity as well as a protection against pre-purchase opening and refilling of discarded packs.

In pharmaceutical packaging the use of holography on metalized blister packs provides similar protection by delivering a secure method of verifying that tablets or capsules within the blister container are real and in pristine condition.

Shrink sleeves are a further indication of the success of embossed metalized foils as they can be combined with the sophisticated shrink sleeves (Figure 7.16 shows how holography can be combined with the tamper evidence of a shrink sleeve) found in use on whisky, brandy and other high value spirit bottles. Here they deliver a surety that a bottle has

**Figure 7.16 -** shows how holography can be combined with the tamper evidence of a shrink sleeve

not been refilled, diluted or the product counterfeited.

Finally, when extended to the protection of heat seal lidding films, the process also provides similar protection to products carried in blow molded bottles and tubs such as lubricating oils, pharmaceutical tablets and other vulnerable products that require induction heat sealing. In these instances it may be necessary to further barrier coat the film to ensure that cross contamination of the product is not an issue.

## INTRODUCING HOLOGRAPHIC EMBOSSING DIRECTLY ONTO PACKAGING COMPONENTS; PLASTICS, FLEXIBLE FILMS AND METALS

In certain situations it is possible to emboss holographic images directly onto packaging components such as clear reflective plastic cases and flexible packaging films as well as metal containers that are coated in a receptive lacquer that is capable of taking a finely embossed image and diffracting the incoming light in such a way that image becomes kinetic.

The process of creating a holographic image in

clear plastic (Figure 7.17) is achieved by creating a specially embossed shim inserted into the injection molding tooling die used to produce each plastic component. The high pressure required to inject the material into the die results in the holographic pattern being transferred to the component during the molding process.

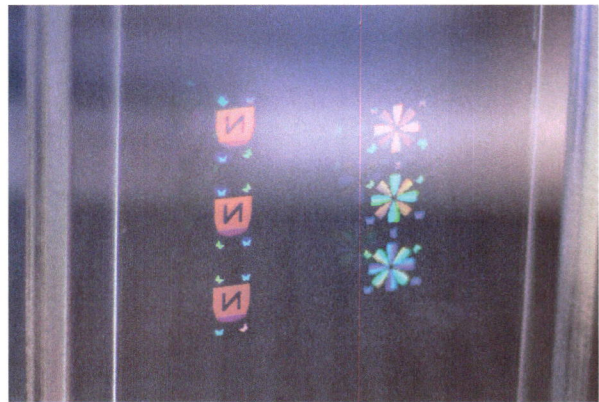

**Figure 7.17 -** Shows the addition of a holographic image directly on to an injection molded clear plastic surface

**Figure 7.18 -** Shows how a nano-holographic structure can be used to secure clear packaging films

A similar process is undertaken to deliver an embossed holographic design to metal cans and containers where a lacquering process is used to coat the base metal after it has been printed and the coating exposed to engraved metal dies that implant the holographic image into the material creating the desired decorative/security effect.

Finally, work has recently been undertaken that moves away from the need for embossing tools to create holographic effects on clear flexible packaging films.

The images produce vibrant colors and seemingly moving images which allow observers to quickly identify the authenticity of a product to which an optical marker is affixed.

The technology works by using nano-scale 'hole' structures - smaller than the wavelength of light - to capture ambient light using ultra high optical efficiency and high pixel density. When such technology is applied to flexible wrapping or laminating film a bright background holographic image is produced that can quickly identify the origin of the material as secure.

## PHOTOPOLYMER HOLOGRAMS

Photopolymer holograms depend on reflectance rather than refraction, (which is the process that delivers kinetic color movement for embossed holographic images). Photopolymer holograms are created in a photosensitive emulsion that is coated onto film in a similar manner to traditional photographic processing.

Reflection holography offers true color reproduction and all round 3D viewing when the device is tilted. The reflection process depends on highly complex chemistry, origination tooling in the form of special lasers, skill in setting up the holographic 'shoot' and the design of a suitably functioning device.

One of the main benefits of reflection holography is that there are only a small number of companies equipped to supply it as a security technology and a very limited number of suppliers of the base film used to reproduce the holographic image, which is exposed and 'printed' into the light-sensitive film coating in the same manner as traditional photographic films were printed before digital cameras were introduced.

This means that each image is exposed to the origination material, which is held on a clear photographic plate that is illuminated by a number of laser beams, and reproduced in a step and repeat process.

**Figure 7.19 -** shows the vibrant color of a film based reflectance hologram

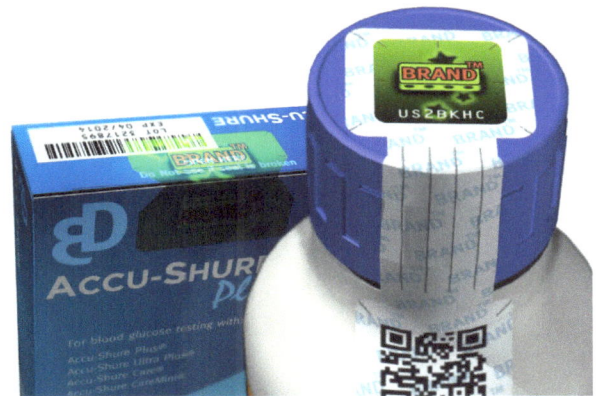

**Figure 7.20 -** adding a clear unique reference to a photopolymer hologram

This allows each image to be superimposed with a further computer generated 'variable' image such as a number or other variable data (Figure 7.20 shows a clear unique reference added to each photopolymer hologram during the run of the original film) that can be used to create unique references that are able to be identified later and traced back to the original piece of film. In the illustration above the security process is enhanced further by the addition of QR code that carries the same reference as the hologram on the tamper evident label construction affixed to the bottle and also to the carton behind.

Since every image is in fact unique and offers a true three dimensional view from each direction (front, rear, left, right) it is possible to introduce features into the design that can only be viewed from the front, rear, left side or right side of the image. This is a powerful anti-counterfeiting and authentication feature that is not available in any other holographic process.

When combined with serialization (next chapter) the process offers the highest levels of print-based security that exist at the moment.

**Figure 7.21 -** Illustrates the combination of traditional security print in a tamper evident label with the attributes of overt holographic security

The photopolymer holographic label (Figure 7.21) shows how the process can be combined with high security printing operations to offer both tamper evidence and authentication protection to a very high value computer component carton.

## SOURCING SECURITY HOLOGRAMS

Holograms can be regarded as one of the most popular methods of securing labels and packaging from the unwanted attention of lawbreakers who may wish to make monetary gain by copying or diverting a brand. As has been revealed in this chapter, holograms are also useful in securing product against tampering and refilling fraud and if they carry further identification features in the form of track & trace numbering they fulfil all the security elements needed to protect and identify non-conforming product from fakes and adulterated replicas.

Because of their undoubted success in this field, holograms themselves inevitably attract the attention of counterfeiters and the sheer number of suppliers of embossed holography worldwide means that fakers are able to source similar if not exact copies of metalized DOVIDs easily via the internet. Indeed, just Googling the term 'security hologram suppliers' reveals 340,000 listings!

Many of these listings, and probably 90 per cent of them, are suppliers and distributors of stock holographic decals and foils that carry various misleading messages such as 'genuine', 'OK', 'authentic' and the like. Since anyone can purchase materials from such sites and then use these to pass off fake product as genuine, bona-fide suppliers need to offer much more secure designs that identify the brand owner and the product, as well as providing a traceable identification number for further security.

If designed and produced competently, a genuine security hologram is easy to identify when compared to a fake device. This process may not be so easy for the public to initially comprehend, but counterfeit holograms can be recognized straightforwardly by inspection teams that have been trained in distinguishing fake from real through just a few well-chosen identification features.

**Note:** To make the process of identifying fake holographic film suppliers easier for the industry, a register of bespoke security hologram designs was created in 1994 and this is policed by the members of the International Hologram Manufacturers Association (IHMA).

Members of this association are pledged to check with the register before taking on any new work to establish if the design they are being asked to produce already exists. The register can't 'police' the activities of non-members, but over the years it has contributed to the detection of copied holograms and even to the apprehension and prosecution of the criminals behind them.

Therefore there is a responsibility for everyone involved in this area of security packaging and labeling to ensure that they supply or purchase from bona-fide, trusted sources and they carry out due diligence before they enter into new supply arrangements.

If you are a packaging supplier or holographic foil manufacturer this means checking that a brand owner is who he says he is before proceeding. Cases where criminals pose as recognized brand owners, wishing to purchase 'genuine' packaging and labeling supplies, for use in counterfeit scams are regularly being uncovered, so it's not just a case of caveat emptor or buyer beware – it's 'supplier beware' too.

# Chapter 8

---

# Adding intelligence to labels – barcode, RFID and beyond

---

The word 'intelligence', often used as a prefix to labels and also packaging, is rather a misnomer since true intelligence requires thought, analysis and intellectual processing. However, intelligent labels and smart packaging are terms we all recognize and associate with systems and devices on packaging that are designed to deliver valuable additional product information to us at point of purchase or point of use.

---

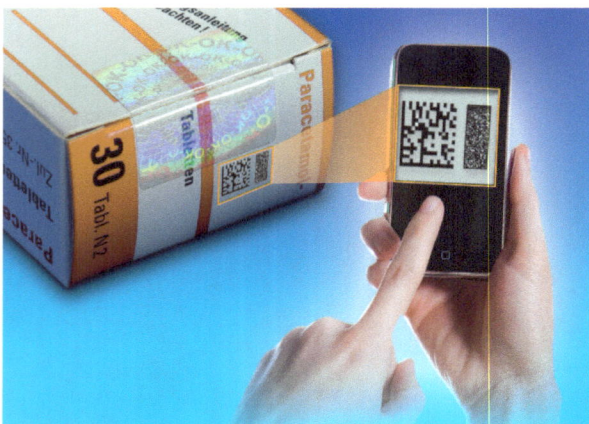

**Figure 8.1 -** Smart packaging is designed to provide additional useful information on a product's status such as its authenticity. Note the de-metalized holographic tamper evident seal

This additional useful information may be acquired visually as a color change indicator for instance, or for more complex information exchange the use of a smart phone may be required.

Labels that react to environmental changes are mostly encountered in the food and drink sector, although color changes can also provide useful indications of product condition for some pharmaceuticals and electronic components too.

These labels inform about temperature change and advisable serving conditions, as well as ripeness condition and in some instances they can delay the effects of product deterioration such as in fresh packed meat and fish.

Most products carry some sort of code. This code can be simple or complex, dependent upon how much data needs to be carried and how important such information is to those in the supply chain and beyond.

In cases where data is used to deliver more complex messages - that may be visually read – this

**Figure 8.2 -** Shows the use of barcodes for tracking individual packs. Note the additional booklet label

will require the aid of a barcode scanner, smart phone or tablet computer.

Product security is a process rather than an isolated event and in order to deliver a safe outcome it is necessary to capture status information at choke points within the supply chain as well as at the location of purchase and up to the final point of consumption.

Status information may include some, or maybe all of the following:

- What is the 'sell-by' and 'use by' date for the product?
- Where was the product manufactured and which production batch did it come from?
- Does the product carry a unique identification number and if so is the product still in the correct distribution channel?
- Has the product been exposed to environmental factors that are outside a safe temperature range?
- Is the product still fresh or has it been exposed to unsafe conditions during transit?
- Is the product authentic and has it remained unopened before use?

## LABELS AND PACKAGING THAT INFORMS ON ENVIRONMENTAL ASPECTS THROUGH VISUAL PROCESSING

As mentioned previously, unwanted or unsafe environmental exposure can be reported visually with a color change. Likewise, the best time to consume a product can also be communicated visually with a color change ink as can the best serving temperature for hot and cold drinks.

**Figure 8.3 -** Color change indicators provide information on a product's freshness and best use date

It is fairly commonplace now to see color change inks on beer labels that change from clear to blue when the correct levels of coolness are reached before serving. Also warnings can be displayed on hot beverages such as coffee and tea that caution that the beverage may be too hot to consume and needs to cool further before drinking.

Other popular applications for such informative labels include ripeness indicators that 'sense' a change in the chemical compounds surrounding fresh fruit in a clear PE⁻ carton and indicate where the fruit is firm, juicy or overripe.

In meat products the ingress of oxygen into containers sealed or flushed with nitrogen can lead to deterioration in the product, making it unsafe or unappetizing to eat. Such labels are useful to warn shop staff as well as consumers that such foodstuffs are past their best and need to be taken off display or thrown away.

Labels can also be used to accurately inform of

water contamination in the field of consumer electronic goods and components.

Water contamination is a common occurrence in the FMCG electronics sector. Products such as cell phones, tablet computers and such can accidentally be dropped into water or exposed to rain or snow during use. This invalidates their guarantee, but repairers may not be aware when a product is returned for service that water ingress is the cause of the malfunction.

Often it is in the interest of the user not to disclose that a broken cell phone has been dropped into a drink or a sink full of water by mistake.

Water contamination indicators (Figure 8.4 ) are supplied in label form and irreversibly change colour alerting to the fact that water damage has occurred. This technology is useful to both the user and the original equipment supplier since it prevents misunderstandings occurring when malfunctioning products are returned for replacement or repair.

**Figure 8.4 -** A product returned for repair or replacement to the brand owner may involve undisclosed water damage. The brand owner can confirm this through the use of water indication labels that change color or flash up a message

Most of the above illustrations apply to safety and provide advice on the recommended best conditions and time to consume a product. Whist not considered directly as anti-counterfeit solutions in their own right;

they add a degree of complexity to the packaging and therefore act as a barrier that counterfeiters can find obstructive if they are unable to obtain the same degree of quality information delivered by these proprietary devices.

## LABELS AND PACKAGING THAT DELIVER 'QUALITATIVE' STATUS DATA IN DIGITAL CODING FORMAT

Unwanted or unsafe product condition can be communicated through color change chemistry that requires the use of specially formulated inks that change their state dependent upon exposure to pre-set environmental conditions.

These solutions, however, can only inform on the current status of a product at any specific moment is time. They cannot identify other critical information that a user or supply chain manager would find useful, such as whether the product packaging confirms authenticity or that the correct distribution path has been taken which would confirm that diversion or parallel importing had not taken place.

Furthermore, data on the source of the product, together with manufacturing information such as lot number, serial number and expiry date are also essential pieces of information that need to be acquired at various points in the supply chain.

Very basic coding systems such as those that only provide an expiry date and lot number or product reference still require protection from alteration fraud since once a product has passed its sell by date it loses value. Therefore some means of preventing alteration or identifying that alteration has taken place is necessary.

This may involve over-laminating with a clear destructible label or debossing using a hard metal die to impress the data into the carton substrate (Figure 8.5).

Alternative approaches could involve the use of hot or cold stamping foil to apply the variable information to the label or pack at the same time as pricing or the Universal Product Code (UPC) is added. A more practical approach for large batches of product is to pre-print this information.

Many producers choose to add this basic data through the use of inkjet printers or laser marking

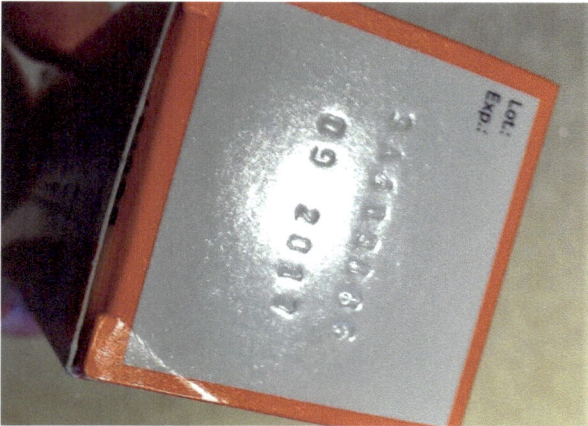

**Figure 8.5 -** Shows how alteration of expiry data on a carton may be prevented by the use of hard debossing

systems. Whatever process is chosen there is a basic requirement to ensure that alterations cannot easily be made, otherwise product integrity may be compromised.

Alteration protection may involve lamination, the overlay of a de-metalized holographic label or the addition of a clear optically variable varnish.

Whilst it is commonplace to see the UPC barcode on many products, this code is static and only applies to a specific product type, size and other common attribute such as color or style. A fixed code will always be displayed on every item through the range produced.

In order to identify a specific item within this range it is necessary to provide every article with a unique reference number so that it may be identifiable and thereby traceable. We see numbers on banknotes, passports and identity documents so it should not be out of the ordinary for us to see them appearing on products too, especially as we move further towards the internet of things (IoT ).

Barcodes are useful since they allow us to acquire relatively long and complex reference numbers that may contain letters and symbols as well as numbers. Such codes provide almost instant machine readability when placed in front of a barcode reader or smart phone camera.

The main driver in the process of tracking and tracing products in this way is the ubiquitous cell phone. It should be recognized at this point that track & trace systems in isolation do NOT deliver authentication. However they can be modified to provide both real-time supply chain information and product verification if designed correctly.

First, let's look at some interesting fairly recent history. With the introduction of digital telecommunications systems in the 1990s it was possible for users to send each other text messages, given that they were equipped with the necessary 2G phones and a suitable key pad.

Very basic authentication systems grew up that took advantage of this capability and it was possible to 'text' a short message service (SMS) alpha-numeric code to a given telephone number and receive a simple reply confirming or refuting authenticity.

**Figure 8.6 -** Shows the use of a scratch off protected code that can be used to check authenticity. The scratch off feature preserves the integrity of the code to ensure 'one time' use

This process, however, was pretty insecure since a database containing the information on the range of serial numbers used could be hacked. If a counterfeiter purchased several genuine products and took note of their codes it was not difficult to predict how the coding structure worked and then introduce copies of the code or predict future reference numbers that had yet to come into circulation.

The benefit that such codes delivered was the ability to check that a pack or label bearing the code was authentic. This could be achieved via a suitably equipped cell phone with Wireless Application Protocol (WAP) or through a laptop or desktop computer linked to the internet. The latter process often required the keying of a lengthy alpha-numeric code and this process in itself limited the flexibility of the system because keying such extended code often led to keying errors which caused frustration and annoyance.

The process has evolved over time and was refined in order to overcome these obstacles. By the turn of the century structured encrypted codes that were not reliant upon a database for checking had been developed. These codes were randomly generated and sent by the user via text message to a server where they were decrypted and verified, removing the hacking vulnerability. In order to protect such codes from copy attacks they were often hidden beneath scratch panels (Figure 8.6) or perforated opaque over-laminates similar to those used by the banks to advise a new PIN for credit or debit card activation.

In some cases the unique reference was also provided as a barcode so that keying was not required and early applications relied upon the camera within the phone for a communication tool. A picture was taken of the barcode and was sent by multi-media messaging (MMS) to a server where the code was decoded and a return message delivered by SMS confirming or denying authenticity.

Despite some of these drawbacks the process of 'serialization' or adding exclusive identifiers was adopted by many brand owners who recognized the benefits that unique item identification provided, especially within the closed supply chain where it became easier to track and trace items on an individual level.

## QUANTITATIVE + QUALITATIVE = TOTAL PRODUCT VISIBILITY IN TRACK & TRACE

Today, the practice of track and trace has advanced enormously, with complex coding structures being developed that are able to carry sufficient information to embrace all the critical data needed to identify an item, where it originated, when it should be used and whether it is in the correct supply channel, to pinpoint just a few of the attributes such codes deliver.

Sophisticated coding systems using inkjet or laser ablation marking can be mounted on packaging machinery and any number of ancillary devices that are required for fast filling, sealing, lidding, and capping. Running such systems requires a database-driven encoding process that delivers the required information at packaging line speeds, often onto a curved surface that can deform the code significantly unless compensated for in software. Such software must also drive quality inspection tools to ensure that readable codes are being produced. This requires the use of scanners that read every code produced and ensure that it meets the requirements of the barcode symbology being generated. (Interleaved 2 of 5, Datamatrix, QR Code etc.)

For perfect 'first time read rates' there will be a requirement to meet edge definition standards, print contrast and dimensional stability in every code produced. If these standards are not met then poor reading performance may well nullify any of the expected benefits that automatic code acquisition may deliver to the track and trace environment.

At this juncture it should again be noted that track & trace in itself is not authentication. Whilst track and trace data provides an essential measurement as to the status of a correctly coded product, it can only provide an indication of origin or veracity. To ensure that the code is trustworthy it must be encrypted and contain a unique (ideally) randomized reference that can be verified using the digital tools that are now available, such as cloud computing which can be reached through wi-fi or mobile connections and a connected intelligent (learning) network that works dynamically to identify fraud as it happens in real time.

This whole process is being driven by what are considered to be the five major digital enablers of authentication and product security.

With the introduction of smartphones such as Apple's iPhone and Samsung's Galaxy series amongst others, the ability of such devices to run powerful apps in conjunction with 3G, 4G (and next

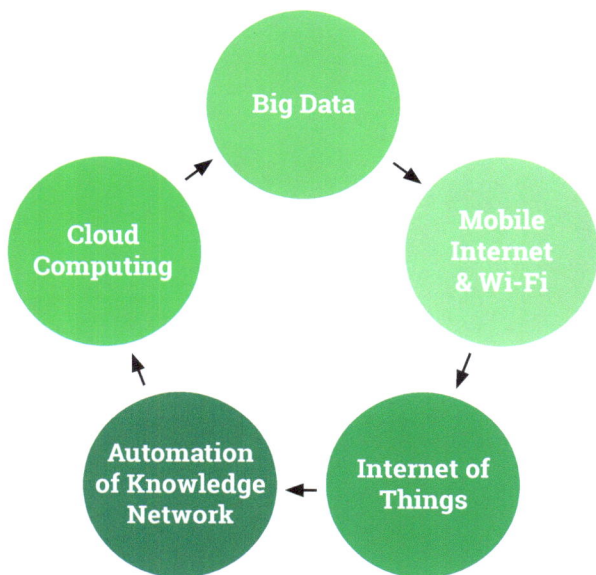

Figure 8.7 - The five major enablers of digital authentication in product security

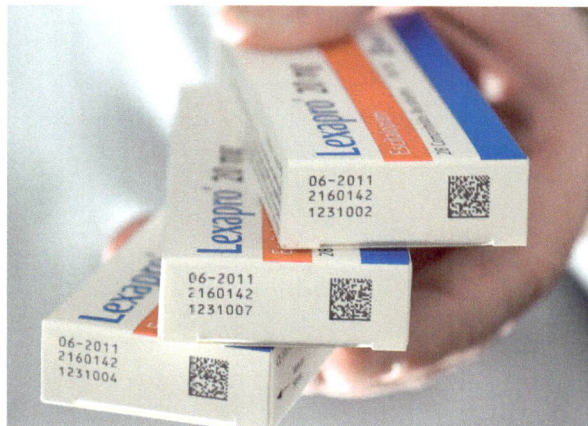

Figure 8.8 - A requirement for prescribed medicines to carry a 2D Datamatrix code and corresponding clear text on every pack

5G) telephony, plus wi-fi connectivity means that more than a quarter of the world's population has access to a barcode reader and the ability to interrogate a product's genuineness and origin through a barcode acquired through the phone's inbuilt camera.

Barcoding is becoming synonymous with security in both pharmaceuticals and tobacco markets as alliances and government directives encourage the use of track, trace and authenticate systems to ensure product integrity and protect against counterfeit products reaching the consumer.

In the pharmaceutical sector in particular there are a number of government mandates already in force; or about to be activated. For instance in Europe the EU Falsified Medicine Directive will come into force in 2018. This means that cartons of prescribed medicines will be required to carry a 2D Datamatrix code (Figure 8.8) and corresponding clear text on every pack to enable serialization and traceability.

Many predict that in the pharmaceutical market protection of the secondary packaging is not enough and primary packaging such as blisters should also carry the code. Taking this a stage further some suppliers of coding systems are providing the ability to code each tablet, capsule and vial with its own protective code. By consolidating codes in this way, and by supplying a master number to every case that leads to subsequent cartons and eventually dose level, a powerful barrier is placed in front of any counterfeiter or perpetrator of fraud such as those that may choose to 'seed' fakes amongst legitimate consignments.

Ideally, to improve efficiency, such systems require the aggregation of the coding structure which is illustrated below and shows how a tertiary case would be marked with its own unique identifier and then how each subsequent pack and sub-pack in the case would be marked so that each individual primary pack carries a hierarchical reference which includes its unique reference/position/identifier amongst others in the consignment.

Another market where track & trace is finding ground is that controlled by the major tobacco companies. Alliances formed by the chief tobacco players in the global market are designed to defend and identify cigarettes which are constantly under counterfeit attack across the globe.

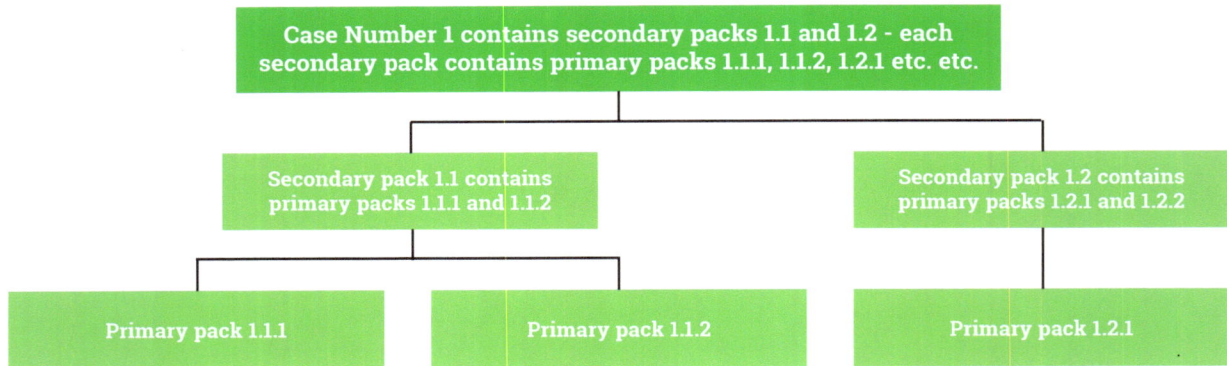

**Figure 8.9 -** Aggregated coding enables each primary pack to be associated with secondary packs and case numbers so all are 'bound up' and inseparable. This approach prevents counterfeits infiltrating the distribution system

The tobacco market has always been an attractive target for counterfeit activity, mainly because of the high excise duty that is levied by governments in an attempt to control consumption and also to raise income to pay for defense, social services etc.

Coding technology offers opportunities for tobacco companies to discover where their products are finally consumed, since the distribution chain can be highly extended and there are a number of opportunities for this to be compromised by inserting fake (or diverted) products at various points within the chain.

The challenge posed for the tobacco industry is that cigarettes are a mass consumer product and universal coding needs to be applied in as many markets as possible, since it is difficult to control the product effectively once it has left the factory. In order to address this weakness the industry has developed a code verification system (CVS).

CVS is a 2D barcode that uses a unique encrypted 12-character number to identify and authenticate a cigarette pack. The number, when linked to a database, can be verified and can be read by a human or by a computer-driven barcode scanner.

By entering the number in the database or scanning the code, a code-verifying computer program will determine whether the code is authentic

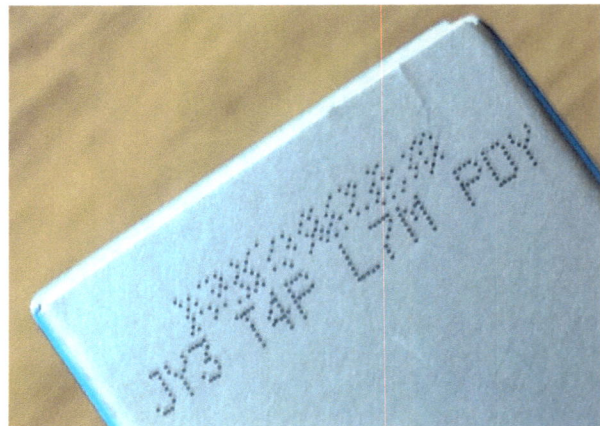

**Figure 8.10 -** The CVS barcode on a cigarette carton allows automated track & trace information to be exchanged and verified

or not. The code has information about the place of manufacture, the machinery, the date and time of production, and the brand.

CVS is a part of the PMI Codentify system. Philip Morris International reports that the application of the codes to product packaging has a minimal impact on the manufacturing process.

PMI has licensed the technology to its three main

competitors: British American Tobacco, Japanese Tobacco, and Imperial Tobacco. Together, they will use and promote this system to governments to ensure a single standard for product verification and identification of non-conforming, tax evading stock.

A similar system is also used for checking the authenticity of cigar boxes. The code is placed on a cigar box before it is sealed and a paper wrapper placed on each cigar.

From these two illustrations it will be appreciated that track & trace in isolation does not automatically deliver satisfactory confirmation of authenticity. A further process is required and this embeds a unique serialized reference to the code that when encrypted supplies the necessary robustness needed to protect both the code and the database from compromise or copy attacks.

It is also important to recognize that such coding technology often exists alongside other more formal methods of overt and covert authentication, since not everyone outside the industry will have access to the code verification system. Therefore it may be observed that holograms, invisible inks and forensic markers (amongst other authentication features) may also be in widespread use in both tobacco and pharmaceutical packaging.

> Readers interested in finding out more about barcodes, code types, code technology, code reading and code verification, might like to study 'Codes and Coding Technology', one of the titles in the Label Academy series of books.

## RANDOMIZATION – CODING THAT IS PRACTICALLY IMPOSSIBLE TO COPY OR DISRUPT

In Chapter 3 it was discovered how randomization provided a useful method of protecting materials from copying and replication assaults.

The combination of randomly distributed colored fibers and planchettes in security label substrates and papers provides the designer with a tool that is practically impossible to copy. This is because to replicate such random occurrences and features

within a substrate is an impractical proposition for any counterfeiter. Mimicking these features is just too time consuming and costly to justify the effort.

The developers of coding systems have also recognized the value of randomization in providing a barrier to the imitation of track, trace and authenticate codes.

In order to capitalize on this attribute a number of security solution providers have introduced various proprietary systems that depend on randomization as a kernel to their serialization and authentication systems.

The process relies on artificial randomization and can embrace unsystematically placed dots, lines or other visually identifiable features such as colored pellets that can be added to a label during the printing process or to the substrate during manufacture.

To demonstrate how artificial randomization works it is necessary to imagine holding, say, six match sticks in your hand. If you throw these lightly into the air they will drop and form a pattern when they finally come to rest on the surface of a table for instance.

Every time you do this a different pattern will be formed. Some match sticks will lie in isolation, others will fall across each other forming a series of intersections. Finally, if the pattern produced is contained within a specific area, all points created by

**Figure 8.11 -** Haphazardly scattered micro wires embedded in a label to generate the random code. This random pattern is 'linked' to a serialization code to assist in processing the data

the crossing of the matchsticks as well as the position of each matchstick can be plotted. It will be noted that each time the process is repeated a different (random) pattern will be formed

Such a procedure will generate millions of different (position) combinations, all of which can be stored in a database as a numeric code. Adding such a code to a label immediately introduces a unique artificial feature that can be authenticated and that is impossible to copy easily. By linking this visual, randomly created code to a serialized barcode that is encrypted, it is possible to provide a useful link between the random visual code and the stored image in the database.

Codes produced in such a way can be read with smartphone apps and verified using software residing in the phone and also in the database (Figure 8.12).

Attempts to copy, say, one code and replicate it many times will be immediately evident since each barcode is unique (as is its accompanying randomized feature) and can be designed to reveal other attributes such as its position within the supply chain, whether it has been read before, and if so how many times.

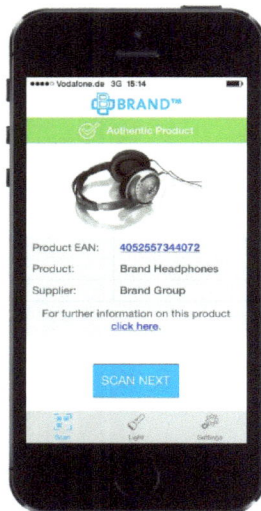

**Figure 8.12 -** Authentication mesages can be delivered through the use of smart phone apps

## RFID AND NFC: AUTHENTICATION SYSTEMS OF THE FUTURE?

This sub-title may seem somewhat confusing since RFID and NFC have already been chosen and proven as ideal authentication systems for incorporating into packaging and labeling.

However this choice is measured against the cost of such systems, which require a complicated infrastructure of readers and also a heavy investment in RFID tags that need to be placed on all the articles within the supply chain that are to be protected from counterfeiting.

RFID is an abbreviation for radio frequency identification and as this title suggests, each tag can be identified by a unique number embedded within the tag that can be acquired through a wireless connection. These 'radio barcodes' consist of a silicon chip, an antennae through which data passes and is stored, and a carrier which is usually a pressure sensitive label.

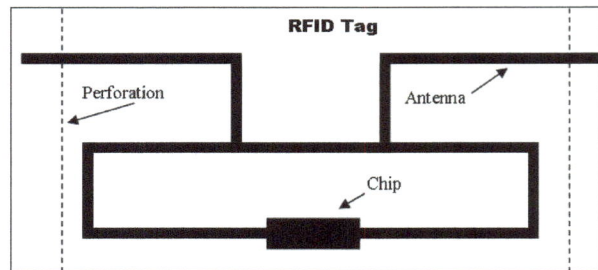

**Figure 8.13 -** Illustrates how tamper evidence can be introduced to an RFID tag

Dependent upon their construction, tags can be read from a distance (Figure 8.13) and can be designed to deliver both reading and writing functionality. Power is delivered to the chip in the tag by the incoming radio signal which also carries a command for the tag to deliver its data payload to the reader.

Such tags can be designed to provide tamper detection by adding a perforation to the label substrate that carries the antennae. If such a label is used as a seal, once the seal is broken so is the

| Short Range (ISO 14443-B) PROXIMITY | Long Range (ISO 15693) VICINITY | Extended Range (ISO 18000-6) |
|---|---|---|

Transport & Event ticketing

Access Control

Library

Industrial

Brand Protection - Customer authentication through cell phone

Laundry

Brand Protection— Supply Chain (hand held reader interrogation)

Retail & Supply Chain management (pallets)

0 cm          20 cm          50 cm          1m          10 m

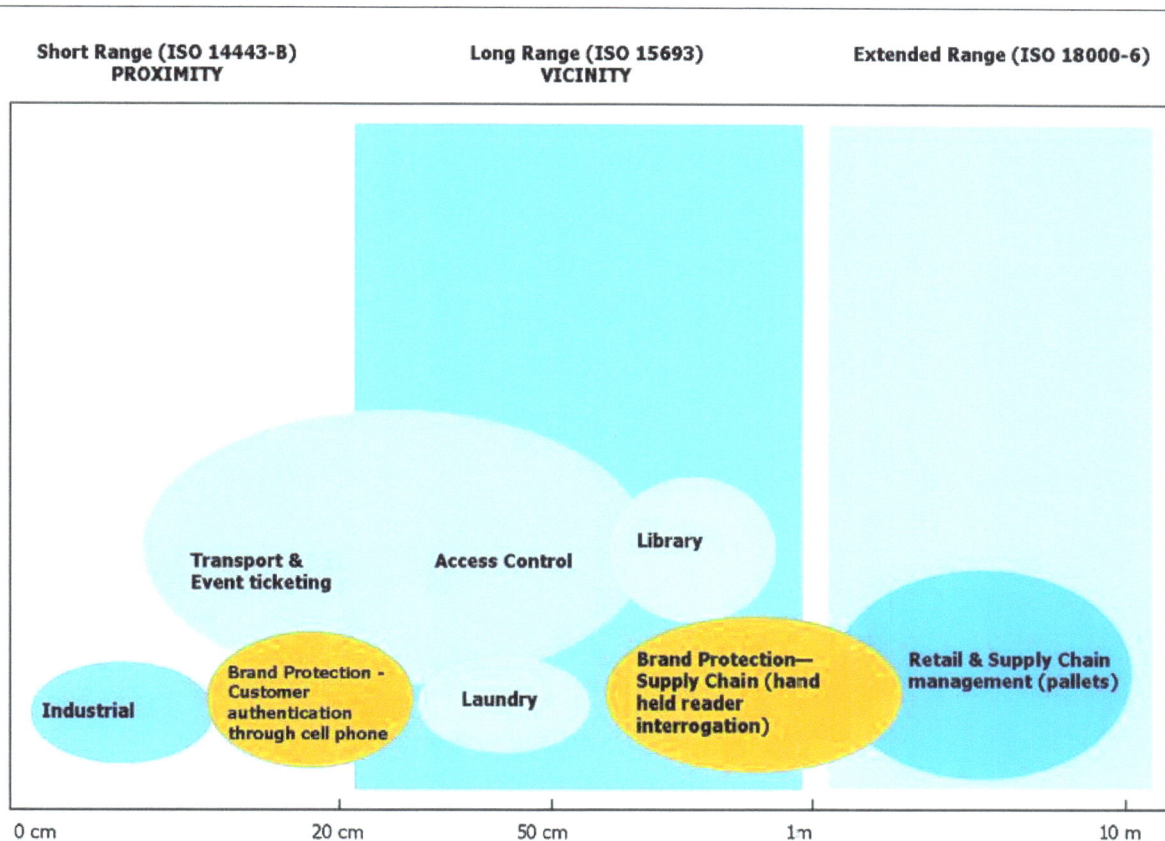

**Figure 8.14 -** Shows the different read ranges and applications for various RFID technologies – note the difference in distance reading between cell phones (NFC) and hand held barcode readers

ability of the tag to communicate.

Even in their most simplified format, such as a radio tag that carries only an identification number, these tags can cost a number of US cents each, even when purchased in large quantities. In small quantities such items require an investment of ten or fifteen cents depending on their format. Therefore until the price of such devices falls significantly their widespread use in brand protection is limited to applications where such an investment is only a small part of the cost of the item carrying the tag.

Therefore such RFID solutions have so far mainly been focused on high price designer goods such as ladies' handbags, shoes and clothing where price

tickets can absorb the cost of such devices and counterfeiting is an ever present threat.

RFID is widely used in a number of other applications such as transport (contactless) ticketing, access control, industrial asset management and logistics.

Manufacturers of RFID tags promote the return on investment (ROI) such devices may provide in brand protection, especially in a well-designed system that leverages on the various attributes of a tag such as providing hands free reading, tamper evidence, authentication, anti-theft protection and supply chain information such as product origin, destination and route to market. These elements when taken in their entirety can all be added to the functionality of a tag, but the incremental additional costs of each performance enhancement feature needs to deliver more in return to make the ROI worthwhile.

For instance EAS, (Electronic Article Surveillance) allows retailers to protect their stock from theft both from display and from within the stockroom. Therefore in such applications combining EAS functionality with authentication will deliver benefits for the retailer and the brand owner who may wish to use RFID as a feature to engage with consumers as well as deliver protection against fake products reaching the consumer through legitimate channels. (See Chapter 1, Figure 1.5)

Nowadays RFID tags can be made with safety features that also protect them from counterfeiters. Just like holograms and color change inks, where RFID tags were useful in brand protection applications there was a demand for fake tags in order to confer authenticity onto phony product.

This weakness was previously considered an inhibitor to market acceptance of RFID for brand protection, but developments that add physically un-clonable functions (PUF's) to the tag have removed this obstacle.

The role of PUF's in protecting RFID tags from counterfeiting involves the recognition of tiny imperfections in the chip carried by the tag. These imperfections act as a material biometric and are written into the tag memory as a self-checking feature.

The evolution of RFID over the last two decades has delivered a trustworthy product that promises to become more widely used in packaging security as manufacturing methods change and printed electronics replaces the more expensive manufacturing processes of the past.

The development of NFC (near field communications) is an important trend in widening the scope for RFID adoption. The growth of this technology, which has a smaller form factor than RFID, but functions in the same manner, has been proven in transport ticketing and in contactless payments.

Contactless payment technology has been previously hindered by a dearth of infrastructure, but

**Figure 8.15 -** The NFC symbol can now be seen on most new issues of credit and debit cards and indicates the contactless functionality of card which requires no PIN for payments of up to £30 or its equivalent in mainland Europe and North America

**Figure 8.16 -** Shows a smart phone NFC authentication process in action. Note NFC symbol on the neck closure. The NFC tag is de-activated on opening, supplying an additional anti-tamper feature

with the recent introduction of NFC applications on Android and now iOS platforms the use of contactless payment methods in stores, coffee shops, newsagents and on public transport has reinvigorated the banks and business as a whole to provide the infrastructure necessary to drive the process further and faster.

Encouraged by the simplicity and speed of the process, as well as the security offered by fingerprint authentication at point of transaction, smartphone users and brand owners are recognizing the power of this solution for brand protection applications.

Again, it is important to highlight the need for delivering a firm 'business case' for NFC in this consumer (not payment) environment. Systems that leverage the attributes of NFC such as secure authentication, tamper evidence and consumer engagement functions will be covered in the next chapter.

# Chapter 9

## Working with the brand owner to enhance and secure the brand

The pressure on brand owners today to grow their market share and face off competitive challenge has never been so strong. For the past two decades brand owners have been drawn in to creating stylish websites that promote their products and offer advice to consumers on how best to benefit from their offerings.

However, with the introduction of tablets and smartphones and the move away from desktop and laptop computing, consumers are now much more active and participative in their habits. Setting up your PC to engage in an Internet session that allowed you to browse a website in order to find guidelines, a recipe, or enter a competition has been replaced by an 'always-on' world that removes the need for setting aside a special time or special place for internet browsing and research.

Such 'internet' activity is now just a matter of tapping a screen and collecting the information you require, as and when you need it. Indeed this trend is supported by the rapid decline in PC use and the replacement of the laptop by a tablet or iPad.

This trend has been driven by the morphing of the ubiquitous cell-phone from a mobile personal telephone into a powerful computing device that offers numerous applications that previously came unbundled as isolated tools in their own right.

Now we can take pictures and create and edit movies on our smartphone as well as run our daily calendar, collect our email and run our finances.

These applications amongst a host of other useful tools are also interchangeable with our tablets.

In fact, our whole daily (and nightly) activity can be scheduled on these devices as we watch live TV, play games, pay our bills and surf the web for shopping or instructions about how best to create our next meal.

Of course, brand owners know that connecting with their prospects and customers through this technology is much preferred to the older traditional methods of marketing such as direct mail and telephone cold call canvassing. As consumers we all seem to prefer the app-based interaction between ourselves and those brand owners we choose to partner with, rather than having to field unwanted solicitations from unwelcome or unknown companies wishing to sell us things we do not need or have no interest in at the present time.

This process has led to the recognition by brand owners of the value of marketing campaigns through social media such as Twitter and Facebook.

With over one billion Facebook users and 500 million Twitter account holders, social media provides brands with clear channels of communication, leading

**Figure 9.1 -** Consumer engagement and authentication is enabled through the use of QR codes and a smartphone app. Note the visual overt security feature in the form of a holographic embedded thread on this label

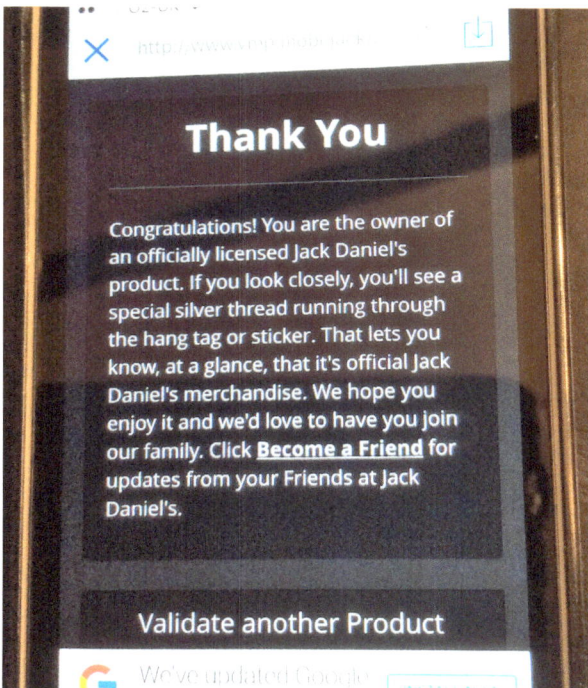

**Figure 9.2 -** Shows the response to an enquiry regarding provenance through the use of a QR code and web generated communication that also invites a 'friending experience'

directly to the customer.

By using these applications, alongside maybe their own specially created smart phone compatible web browsers and other interconnection tools such as barcodes and now NFC tags, such technically savvy brand owners are attempting to infiltrate our social networks and become part of our daily activities through the use of 'like' and 'friend' tabs for us to activate on their web pages.

The most common point of connection for consumers with a brand is the branded pack or label they see on the product they are about to purchase in store or about to use or consume at home.

For want of a better terminology we will call this the 'connected package'. The illustrations (Figure 9.1 and 9.2) provide a direct insight into how this process works and shows a Jack Daniels application (December 2015) that links authenticity checks for the consumer with an orchestrated response, confirming how to check and verify provenance through the security features present on the label and also provide an invitation to become a friend.

## PACKAGING AND LABELING THAT IS RESPONSIVE TO CONSUMER BEHAVIOR, RETAIL EXPERIENCES AND CULTURAL PHENOMENA – MOBILE ACTION CODES

Packaging and labeling has the potential to create much greater value for brand owners, retailers and consumers. The 'connected package' can deliver more value to each participant in the process by becoming part of the digital stream of information, communication and of course the transaction.

The connected package or label becomes more useful and as a result, it weaves itself seamlessly into the lives of consumers and the logistics of distributors and retailers, delivering greater value to all in the process.

In order to bond with the consumer and so that it may deliver extra value to the brand owner and the retailer, the connected package needs a quick and easy to use link that can connect each partner in the process, and this link may be a bar code or an NFC tag as we can see from the illustration (Figure 9.3).

This link could also be an embedded invisible digital watermark placed within the print carried by

| Augmented Reality | Mobile Search |
|---|---|
| QR Codes | Apps |
| Mobile Ads | **Mobile Marketing** | Mobile Site |
| SMS | Location based services | Mobile Search |

**Figure 9.3 -** Mobile marketing provides a number of extra benefits to the process of adding QR codes that invite consumer response

the label or pack and extracted using an app on a smartphone.

Indeed, as has been revealed in the preceding chapters, digital technology in the form of codes in visible, invisible or RFID/NFC format, act as 'triggers' that automatically set off a recognition process if they are activated by a live app when we touch part of the screen on our smartphone or tablet.

These triggers link us then to some form of interactive content that allows us to engage in some way with both the brand owner and the product.

This engagement may be just a simple 'like' action through the brand's Facebook page which was acquired through scanning a QR code, or it may be more complex such as delivering an authentication message and a 'thank you for your purchase' response from the brand owner who may be able to learn more about your location when you scanned the code and compare this data with the knowledge of

where the purchase took place and whether the product was inside or outside the recognized distribution channel.

Engagement may also involve loyalty programs and, of course, before and after sale feedback with discounts being communicated on future purchases and links to other connected products that may attract the customer's attention.

It will be seen from the illustration the importance of gaining initial awareness for the brand owner. This informs the consumer that more information exists that can be acquired and assist with the purchasing decision, maybe swaying the buyer in favor of choosing one brand over another.

Engagement through the label and packaging at this point is critical, and since real estate on these mediums is always at a premium, a 'trigger' that is highly visible and simply activated such as a QR code or logo informing that an NFC tag is present or an Augmented Reality experience is available to share, needs to be present (more about Augmented Reality later).

After initial engagement the consideration process should hopefully lead to conversion or sale and at this point it should be possible to build further brand loyalty and repeat sales if the consumer enjoys the experience and feels that the relationship is permanent (e.g. engagement leads to marriage).

Those readers qualified in the subject of marketing and customer relationship management which is a core function of marketing will be familiar with the 'marketing mix'.

Marketing is the management process responsible for identifying, anticipating and satisfying customer requirements profitably. It involves four core distinct

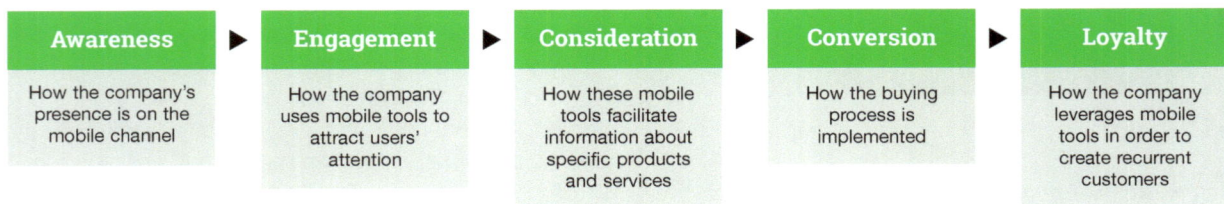

| **Awareness** ▶ | **Engagement** ▶ | **Consideration** ▶ | **Conversion** ▶ | **Loyalty** |
|---|---|---|---|---|
| How the company's presence is on the mobile channel | How the company uses mobile tools to attract users' attention | How these mobile tools facilitate information about specific products and services | How the buying process is implemented | How the company leverages mobile tools in order to create recurrent customers |

**Figure 9.4 -** Mobile marketing as a business strategy is illustrated in Figure 9.5

| Mobile Integration strategy | Level 1 Integration | Level-2 Integration | Level-3 Integration |
|---|---|---|---|
| **Product** | Mobile interaction through information on packaging | Integration of interaction into packaging or labeling NFC/QR Code/AR | Built in mobile feature on core product labeling (for use after packaging is discarded) |
| **Pricing** | Mobile strategy to reduce cost of goods sold | Mobile interaction built into above to create upsell opportunity | Checking product pricing/ channel and authenticity |
| **Place (distribution)** | Mobile strategy to direct customers to nearest point of sale | Mobile as a tool for customers to access point of sales | Location based marketing decisions can be made based on the point of interaction with |
| **Promotion** | Maximise audience reach for awareness & interaction building | Interactive promotions driven by labelling and packaging to build CRM | Seamless engagement to create AIDA. (Attention/ Interest/Desire/Action) |

**Figure 9.5 -** Mobile marketing analysis explored in detail – authentication is a natural progression of this process

disciplines; Product, Pricing, Place and Promotion.

To remain competitive brand owners need to ensure that they place the right product in the right place at the right time and ensure that it is correctly promoted to enable them to maximise their return on investment. This investment will include R&D, product development, manufacturing costs, raw materials, administration, distribution and promotion.

A primary objective of this chapter is to show how the disciplines of good product security can be coordinated with those of labeling and packaging design, as well as exemplary product marking practice.

Mobile marketing, combined with effective authentication tools and interactive labeling and packaging can help immeasurably in delivering a cost effective marketing mix for the brand owner. So much so, that those practicing this wider discipline, together with a structured and strategic approach to security, can ensure that any risks or vulnerabilities (Chapter 1) are satisfactorily addressed.

With the introduction of the ubiquitous smart phone, alongside the ability of these devices to run mini-programs or apps, together with a high definition screen and a camera, there has been a plethora of apps that offer an interactive experience to the engaged consumer.

Although action codes such as QR codes and embedded NFC tags are able to offer quick connectivity with products and deliver pertinent information quickly and efficiently, other methods of providing interactivity also exist.

Augmented Reality applications and associated electro-conductive enabled labels, tags and packaging are able to acquire a trigger and link to interactive web based content without the need for a visible printed code or even an NFC connection.

## INTRODUCING INTERACTIVITY AND AUTHENTICATION FROM AUGMENTED REALITY AND TOUCH-CODE APPLICATIONS

Augmented Reality (AR) is a diluted version of virtual reality. As we know, virtual reality requires the user to wear a helmet so that they can experience the virtual world that is computer animated and delivers the illusion of reality as they interact and become engaged with a computer program.

With augmented reality the user experiences a

new world through the screen of a smartphone or tablet computer with the assistance of the camera and a suite of overlays that are superimposed on the camera's real-time view of the world.

The process of augmented reality can also be delivered through a webcam set up on a PC but the popularity of hand held tablets and smartphones has superseded this technology.

There are a number of apps available through both Android and iOS platforms that deliver an AR experience which can include sound, graphics, GPS data and video, superimposed over the real world view captured by the camera. However each app is specific to its own triggers and consumers need to download a number of these applications if they are to interact with products using a variety of AR configurations.

From a labeling and packaging viewpoint, AR content can be superimposed on real world images such as the packaging itself. With AR no visual trigger is needed since the app 'recognizes' the product through a series of specific graphical points, or even logos that are unique to the brand's packaging or labeling on that product.

In the illustration (Figure 9.6) when the tomato ketchup bottle is 'scanned' using an AR app the label is recognized and instantly comes to life in an animated delivery of a recipe booklet that can be flipped open and viewed page by page through tapping the 'open' button.

Consumers can connect via Facebook with the brand owner's account or enter a competition by selecting the option highlighted by the tomato picture in the bottom left hand corner of Figure 9.6.

One important restriction on most AR applications is that unless these interactive images are stored within the app itself, they must be acquired through a server operating via the phone carrier signal or through Wi-Fi connection.

There are literally hundreds of brands utilizing this technology (amongst them Mondelez, Heinz and Unilever) which is particularly effective when delivered in store as part of the highly competitive purchasing decision. A number of well-known toy brands and auto-motive manufacturers in particular have successfully used this engagement process to

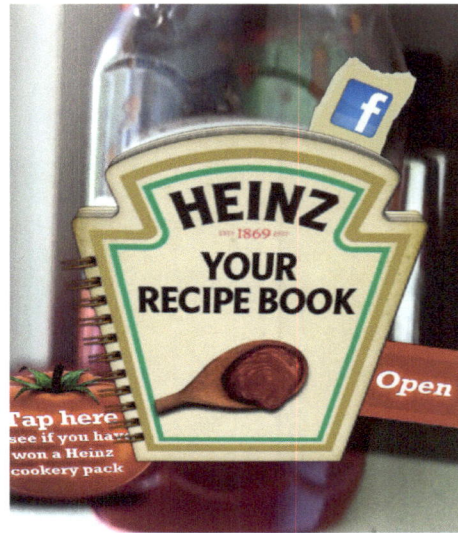

**Figure 9.6** - This Heinz ketchup bottle delivers an AR experience when an app is switched on and the smartphone or tablet is directed towards the bottle label

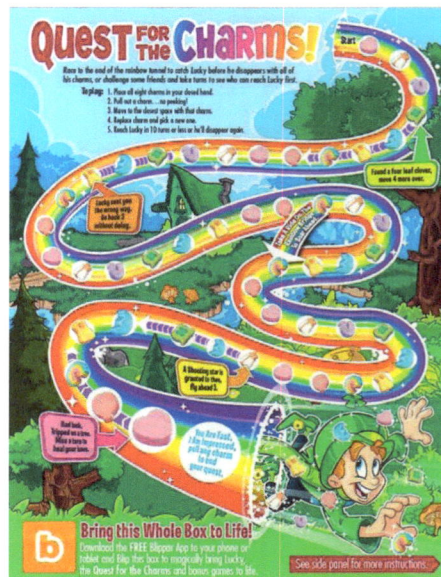

**Figure 9.7** - The orange 'b' logo in the bottom left hand corner of the illustration notifies that a 'Blippar' AR experience is available

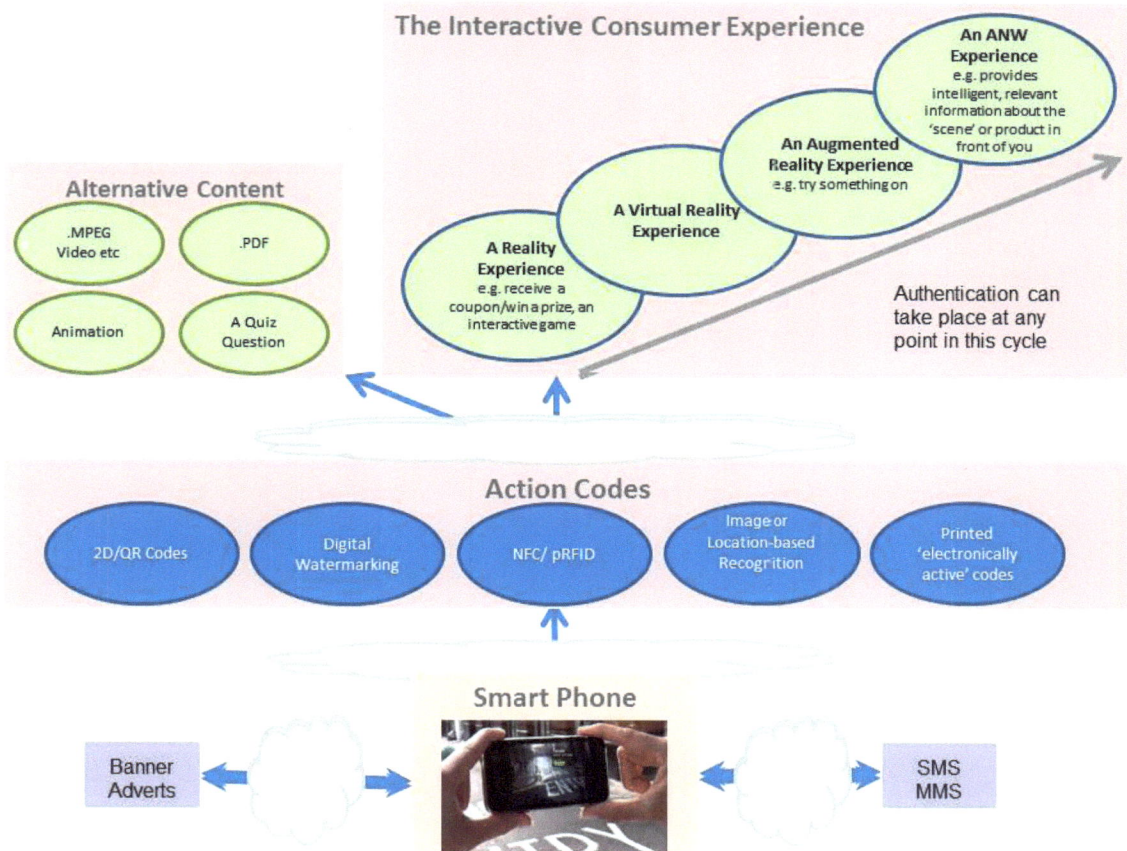

**Figure 9.8** - The illustration above shows the various 'triggers' required to deliver action in the form of video, music and other interactions aimed at engaging with the consumer

increase their chances of winning business against those of their competitors.

Consumers are made aware that an AR experience is available through the use of a graphic on the packaging or product label alerting to the fact that an app is required to provide additional video or audio content.

A further benefit of AR is that it does not expire after the initial viewing and can be modified to deliver a number of alterative experiences that offer different attractive content such as competitions, games and directions well after the initial purchase.

At this juncture readers may well be asking what AR has to do with product security.

Since AR requires highly accurate printing to enable it to operate, such technology is highly sensitive to the quality of the printed image being viewed. Every individual AR image must be enrolled and of course the 'experience' programed before it can be promoted.

Some AR providers use a version of digital watermarking to act as a 'trigger' when a product is scanned. This trigger can also have authentication keys built in to the image that verify its provenance

**Row and column stack up layers**

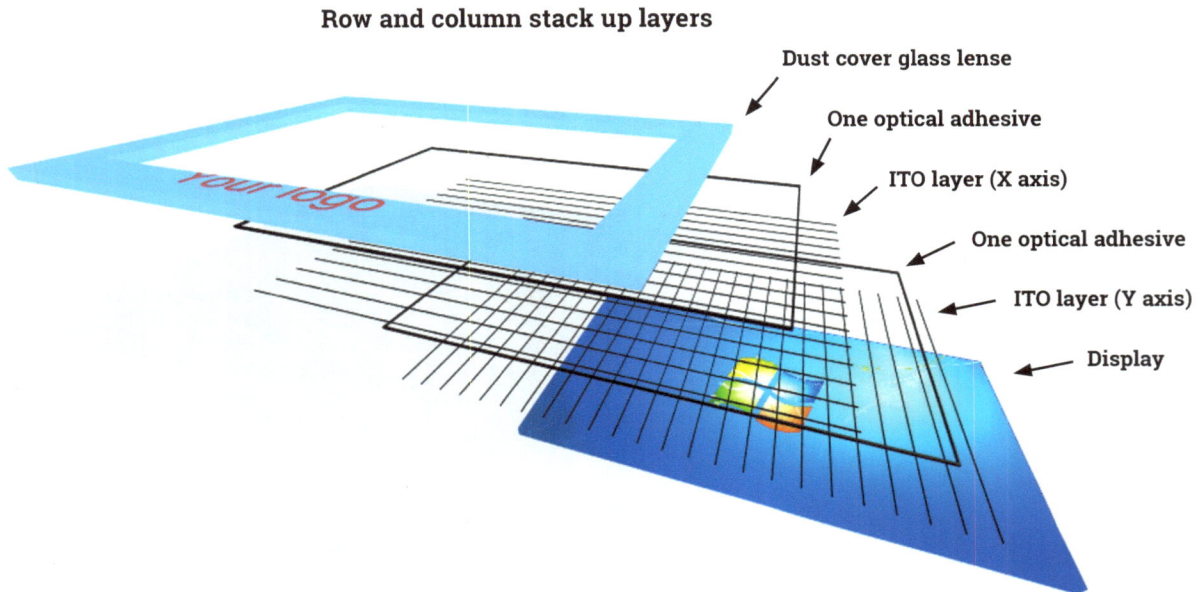

**Figure 9.9 -** Schematic of a capacitive touch screen

and if printed digitally these covert images can be used to identify each unique product.

Combined with a location and the individual fingerprint within the phone or tablet, a brand owner has the ability to detect further critical marketing intelligence from such AR activity. This intelligence can provide further useful information such as in and out of channel information and exact geographical data about where the AR activity was activated.

These actions however must respect privacy concerns and the sensitivity of such data collection operations being possible are still being questioned on ethical grounds.

## PRINTED ELECTRONICS AND CONSUMER ENGAGEMENT

In order to remain competitive with all the other forms of interactive media that exist, print has to continually reinvent itself and the move towards printed electronics is a further illustration of this important trend.

Since both smartphones and tablets use touch sensitive screens it is possible to leverage these capacitive devices as a trigger that will link specially printed labels and packaging to the digital space where an interactive event is stored and delivered through Wi-Fi or 3G/4G telecommunications channels.

The capacitive touch screen on your smartphone or tablet is able to distinguish and sense specific touch location, based on the electrical impulses in a human body, typically the fingertip.

The touchscreen consists of a complex mat of connections that can be activated by the touch of a fingertip on any specific point within the display on the screen. Of course the screen display needs to be programmed to overlay these electronic triggers.

Specially formulated electro-conductive inks can be made that imitate the electrical impulses present in the human fingerprint.

These inks can then be printed in specific patterns onto a label or pack so that they trigger the software in an app to recognize and respond to a pre-programed pattern present in the printed pieces.

Of course any printed code such as this would interfere with the other graphics present on the label or pack. However, because it's an electro-conductive ink, this can be placed on the back of the printed piece or underneath a suitably printed opaque top layer.

The subsequent code is activated by opening the app and then laying the interactive piece of print on the screen surface of the tablet or smartphone.

**Figure 9.10 -** Overlaying this hang tag on a capacitive screen delivers a consumer engagement experience through the application of a specially formulated electro-conductive ink

Once this is completed and the app has recognized and verified the presence of the touch code, the print piece can be removed or repositioned for further interactivity to take place. This may involve playing a game by touching various sides of the pack to the screen in a certain order, or it may be used to deliver a message confirming authenticity.

Other versions of consumer engagement are also possible, such as the delivery of a recipe for a meal that includes instructions on how best to serve a brand of food or wine in accordance with the most appropriate cooking methods or trimmings.

Using electro-conductive inks in this way also provides the brand with a method of audibly connecting with sight impaired or dyslectic users since this method of interaction can be configured to deliver audio files that give instructions such as when

and how often to take medication. It is a more practical method of communication for the visually impaired as it does not require the careful manipulation of a piece of print and the coordination of the smart phone or tablet camera.

## FURTHER TRENDS IN PRINTED ELECTRONICS

With the ever decreasing costs associated with printed electronic circuits, transistors and printed polymer photovoltaics (printed thin film batteries) providing power sources in formats that resemble a traditional self-adhesive label, there has been a surge of interest in attention grabbing labels and packaging that provides eye catching printed flashing OLEDS's (organic light emitting diodes).

**Figure 9.11 -** A printed electro-luminescent display seen on a Scotch whisky bottle

Also electro-luminescent inks and connected image changing displays from companies such as E.Ink - who have taken the monotone display used on the Kindle and modified this so that it can be incorporated into packaging - are now coming to market (Figure 9.12).

Whilst these features were never designed to be anti-counterfeit devices in their own right, they provide a robust defense against copy attacks because they are technically challenging and require a high degree of investment in the plant needed to manufacture

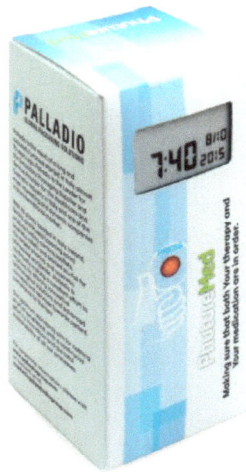

them in economical quantities.

Traditional printed packaging and label formats based on paper, plastics, metal and printing ink are reaching a point where they can develop no further. Space on product packaging is limited and increasing pressures to include more information in order to provide safety advice and nutritional facts and figures alongside instruction and all-important branding and marketing regalia combine to form an impossible challenge to packaging design and functionality.

This 'perfect storm' driven by the need for ever more information, along with increased legislation, especially in the pharmaceutical arena that demands 'total visibility' of the product right through the supply chain, is driving the necessity for visual communication methodology that is not space intrusive.

Technology can provide answers to these demands through the clever amalgamation of miniaturization, printed interactive displays and

**Figure 9.12 -** shows a variable printed electronic display

| | QR Codes | AR/Image Recognition | Electro-conductive Inks | NFC |
|---|---|---|---|---|
| App Needed? | Yes | Yes but customized to EACH provider | Yes | Optional – most IOS/Android phones have NFC tools embedded |
| App launch from background? | No- need to open scanner app (ie QR Reader) | No-must open app first | No- must open app first | Yes- Tap NFC tag from and screen |
| Consumer ease of use | Bit messy | Can be fiddly to first acquire trigger image | Initially, user needs patience to activate feature | Very simple |
| Compatible with curved packs/labels? | Not recommended for placement on acutely curved surface | Yes- but curved surface needs to be 'enrolled' for each application | No | Yes |
| Serialization/Authentication | Yes – either immediately with encryption or through 4G connection with database | No – not generally available but one provider offers this a well as AR | No- not yet but with digitally printed inks a coding structure could be developed | Yes – either immediately with encryption or through 4G connection with database |
| Sensor integration | No | No | No | Yes |
| Cost driver | Low cost in coding but needs software to drive it | Developing and enrolling each AR trigger. After that minimal cost involved | Developing and enrolling each AR trigger. Ink cost | Tag cost needs justification against benefits |

**Figure 9.13 -** The various attributes of coding technologies explored

intrinsic communication methods that provide an intelligent link through the world wide web. This link, together with ubiquitous hand held devices such as smartphones and tablet computers that offer simultaneous interaction and authentication assurances, is fast becoming a favored method of fighting counterfeiting and other product related crime.

Indeed, the very assets that the counterfeiter relies upon to assist in crime - the internet for marketing fake products; and computers, scanners and printers for the production of bogus packaging and labeling - are being turned against the criminal.

## SMART INTERNET TOOLS THAT CAN TRACK DOWN FAKE PRODUCTS ON THE WORLD WIDE WEB

All of today's brand owners that have identified counterfeiting as an important diverter of revenue from their bottom line performance, recognize that it is not only essential to protect their assets that are distributed through traditional routes such as retail stores, but also they need to safeguard their brands on the internet.

Recent statistics reveal that one in six products purchased online is a counterfeit. Since there is a high degree of anonymity present when sellers promote fake products through online markets such as eBay and Alibaba, such platforms are popular ways and means of reaching new customers with a minimum of risk.

Setting up fake websites that mimic those of legitimate brand owners is also a popular tactic since any number of misleading domain names can be used in order to direct possible purchasers to a bogus website that acts as if it belongs to a genuine brand.

The trick for brand owners is to react to these threats via social networking in order to educate customers and alert them to the dangers of purchasing fake product.

For instance LGG® created a Facebook page that features videos and photos that help consumers identify counterfeit product and alert them to popular scams. In addition, the brand also answers questions from customers and offers support with identifying sites as counterfeit or genuine.

Brand owners often deploy special software that enables them to search the internet for brand logos or instances of text references to their brand names and other intellectual property they may own, such as marketing sign offs they may use to describe their products. A popular sign off such as the McDonald's 'I'm loving it' can easily be recognised using the right search tools.

These automated search tools once activated with a reference to a brand logo or sign off, can provide robotic searches of the internet and analyse every reference made on the web to each specific targeted phrase or logo.

Through actively controlling and monitoring internet content in this way it is possible to quickly identify and take down offending websites and prosecute persistent offenders.

Many of the leading providers of security labeling and packaging solutions that are designed to detect and deter counterfeiting, employ such internet monitoring tools as part of their service to their clients.

The need to provide a holistic service that delivers 'a one stop shop' approach to brand protection is an important part of any marketing strategy for businesses operating in this space.

# Chapter 10

# Managing security issues as they affect products in the print business

The challenges relating to product security have always been present. From the earliest of times there has been a need to protect products from copying, dilution and tampering.

For instance, the Sumerians developed a very early form of labeling using clay tablets that carried the instructions for product use of goods that were stored and traded in baskets and amphorae (clay jars). By passing a piece of thin rope through the clay tablet they were able to create a basic form of tamper protection and guard against pilfering and copy attacks since each 'label' was individually crafted, fixed to the sealed closure (which was wax) and this acted as an early form of branding and quality assurance.

The threat of counterfeiting has persisted over the millennia and this is always driven by substituting inferior, less costly ingredients, materials or components that provide the fakers with their profit and incentive to continue the crime.

In Germany a new quality initiative was introduced in 1516. By insisting that the 'only ingredients used for the brewing of beer must be barley, hops and water' it ensured the quality of the product and threatened legal sanction against transgressors hundreds of years before anybody had heard of consumer protection laws.

Appropriately, one of the key reasons for the introduction of 'Reinheitsgebot' - to give the law its

**Figure 10.1 -** An early form of labeling and communicating a product's ingredients was developed in 2000 BC

official German name - was to protect beer consumers. Five hundred years ago water supplies were often polluted so people drank beer, often in vast quantities, to keep thirst at bay.

By ensuring beer contained only high-quality products the law protected the public from poor standard, and potentially lethal, beverages.

Reinheitsgebot also brought about standardization in production well ahead of its time. Foreign brewers who wanted to enter the local market also had to stick to the law and so its influence began to spread far and wide.

Quality and safety measures to protect against the risks associated with product related crime have evolved steadily over the years culminating in the introduction of many of the more sophisticated procedures today such as the BSI (British Standards Institute), UL (Underwriters Laboratory) and others which are overseen by the ISO (International Standards Organization).

The search for quality has also permeated the world of print and for the last quarter century the industry has striven towards systems and procedures that measure and test quality across the print shop floor and also into the administration areas of the business.

## THE IMPORTANCE OF QUALITY MANAGEMENT AND PROCEDURES WHEN PRODUCING SECURITY RELATED LABELS AND PACKAGING

The quality of labels and packaging used to protect and contain consumer goods offers a useful first indicator of product provenance. Conversely it can also be a guide to fake products as counterfeiters often make noticeable mistakes such as misspellings and poor color matching when copying packaging and labels.

Of course, poor quality fakes are pretty easily identified so counterfeiters now take more care over their attempts to copy such items more carefully.

Since quality is an important factor in deciding the provenance of security related packaging and labels, it stands to reason that every production batch of these items must be identical. By setting strictly controlled production guidelines and quality

procedures it should be possible for the producer of such components to deliver identical copies during every production and subsequent production runs.

Managing quality will require a careful control of both materials and press settings in order to match different tranches in production and an agreed reference sample should be used to achieve this objective.

Since the reference sample will also carry all of the authentication 'security' devices specified by the client it follows that these should all be strictly monitored as well

A word of warning is needed here.

The higher the number of security features included within the label or packaging design, the more complex the quality monitoring process becomes. This is because where there is a requirement for an authentication or other security feature (or features) to be present they must be existent on every production piece of packaging or labeling, otherwise the brand owners investment in a product protection system is nullified.

Therefore a careful balancing exercise is required in order to ensure that wastage is kept to a minimum during the set-up and make ready of each additional security feature present. This will also include on-line automated inspection procedures as well as manual quality controls such as sampling and viewing.

To assist in this process there are numerous on-line and offline auto-inspection systems for controlling color, checking print character and logo quality and monitoring invisible inks, hologram and foil position quality and bar code/clear coding performance. There are organizations that can help with advice and training in this area. In Europe, Intergraf has a section related to security printing quality and in the USA, NASPO (North American Security Products Association) offers similar advice and training.

## PHYSICAL SECURITY REQUIRED IN PRODUCTION PLANTS

It is also important for producers of security related packaging and labeling to recognize the need to protect all the materials and waste within their production plant. Established suppliers in this market operate from secure premises that are guarded 24/7 and ensure that all their staff working within the areas producing security product are trustworthy and carry suitable identification credentials within the plant. Access control must be monitored and staff logged into and out of sensitive areas within the production and storage areas.

It is also necessary to secure and monitor the perimeter of the plant and install CCTV in order to observe and record movements of staff and visitors.

General guidance in this area would also include securing all origination files and platemaking/imaging equipment, and overseeing the destruction (or secure storage) of printing plates at the completion of each job to ensure that they cannot be stolen or reused unofficially.

Likewise it is also good practice to audit and inspect those suppliers tasked with providing security inks and materials such as watermarked pressure sensitive papers and holographic stamping foils. Their production, storage and delivery systems must also be protected against the leakage of sensitive components that could be useful to those wanting to compromise or copy legitimate authentic labels or packaging.

Established practice in the security printing environment is to create a 'chain of custody' from the earliest moment in the supply chain where a compromise could occur. This means accounting for every foot of raw material through to every hologram on an individual basis to ensure that everything in the process is accounted for and recorded.

Records that are capable of being audited should register the length of each reel (or a counted number of sheets) in the production run together with an itemized account of all waste which should either be shredded immediately or placed in locked containers until it can be securely disposed of.

Numbering (or uniquely identifying each label or tag) is an important part of this process for more valuable items such as certificates of authenticity or labels that will be used as part of a controlled distribution of products that are to be protected by an authentication program.

In cases like this the boxes containing the finished labels should be security sealed with tamper evident

**Figure 10.2 -** Numbering labels and tags is an important part of inventory control - as well as an authentication tool

tape and a record placed on each box recording or listing the numbers of the labels within the box.

The above pointers are provided as a general overall guide to securing a production plant that produces labels and packaging for applications that address product related crime threats. For more complete guidance those interested in this topic should consult with their chosen quality assurance provider/auditor.

Of course the importance of traceability should also extend along the full length of the distribution chain so that the 'Chain of Custody' is unbroken and authentication and product pedigree is preserved though each hand-over point in the sequence of events that take place right up to the point of final use by the consumer.

## SECURITY IS A PROCESS, NOT AN END STATE

All security systems (whether conventional or print related) should be capable of deterring and defending against attack. If circumvented then security should be modified accordingly in order to mitigate against future attacks. This continuous process requires a high degree of attention and also a firm plan that includes a clear path of what steps are required for

**The primary purpose of authentication is to safeguard the legitimate supply chain against penetration of counterfeits and create a 'chain of custody'**

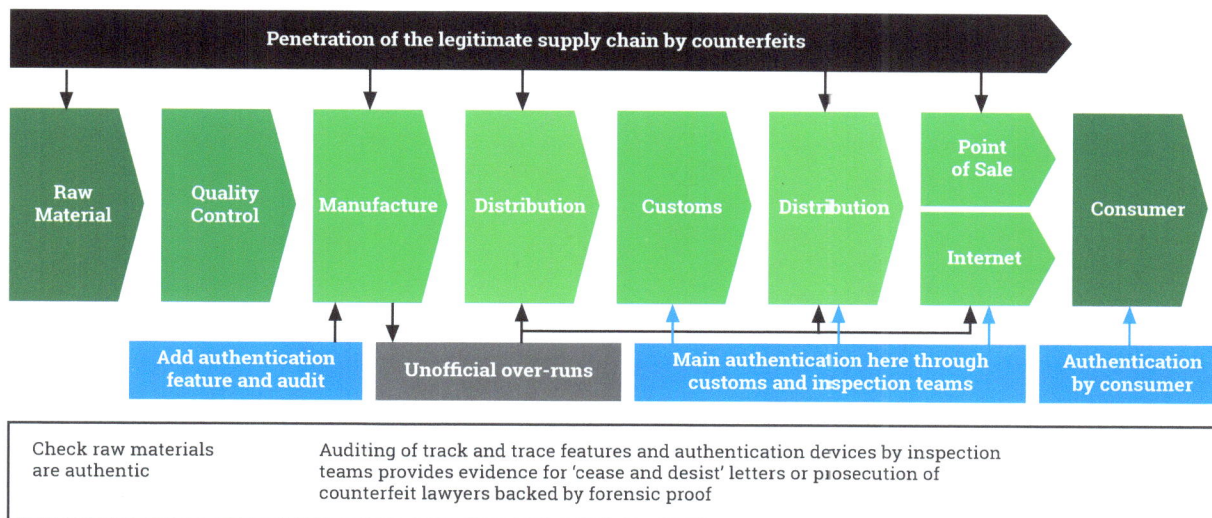

Penetration of the legitimate supply chain by counterfeits

| Raw Material | Quality Control | Manufacture | Distribution | Customs | Distribution | Point of Sale / Internet | Consumer |

| Add authentication feature and audit | Unofficial over-runs | Main authentication here through customs and inspection teams | Authentication by consumer |

| Check raw materials are authentic | Auditing of track and trace features and authentication devices by inspection teams provides evidence for 'cease and desist' letters or prosecution of counterfeit lawyers backed by forensic proof |

**Figure 10.3 -** The process of securing the supply chain

future upward migration should this be necessary.

There are four steps involved in the security process: assessment, protection, detection, and response. Assessment is the leader of the process because it helps to prepare for the remaining three components. Assessment deals with any policies, procedures, laws, regulations, budgeting, and managerial duties including technical evaluation of the security status and risk involved to each product/ package/label in the brand portfolio.

A failure to account for any of these can compromise the flow of other operations in the process.

Next is protection. Protection is when countermeasures are applied to help limit the likelihood of compromise to the product occurring. This will involve the introduction of security features that protect against the threats of counterfeiting, tampering, dilution etc.

Protection and prevention may both be used interchangeably.

Detection comes after protect/prevention in the process. Detection is when policy violations or security incidents are identified. This will occur during a routine inspection process in the field or through a customer complaint or other related intelligence such as notification from border protection (customs) agencies.

The final step in this process is response. Response can be defined as the process of validating the fruits of detection and taking steps to remediate IP infringements such as counterfeiting or diversion. This will often involve the use of legal co-operation and the pursuit of any culprits through the courts.

It is at this point in the process that the careful choice of security features will become beneficial since third parties such as customs, investigators and legal teams will find it more useful to communicate the procedures required to correctly identify a counterfeit via a given security feature on the label than if they have to inspect each product individually and follow instructions for identifying specific visual or weight deviations in quality.

Likewise expert witnesses will find it easier to compare (in courts of law) such security features that are present on the original products with those that are not present or imperfect on the fake goods.

A detailed illustration (Figure 10.4) provides a more comprehensive view of how the process functions.

The appropriate checking level for consumers, specialists and forensic examiners is provided together with what is considered to be the applicable security feature(s) suitable for inclusion in brand protection programs. The verification tools necessary to reach a decision at each level are also included.

It should be noted that the three basic levels of authentication - overt, covert and forensic (see Chapter 2) - can be converted into practice through the suggested security features at each of the three levels within the pyramid. There are also suggestions regarding the types of verification/authentication tooling required for each level.

By following the guidance offered by this illustration it is possible to select any number of security features and deploy these across the various levels of inspection/authentication required. Again, don't forget that the more security features selected the higher the wastage will be on the final print run and the more difficult it will be to control quality on each additional feature.

## Authentication strategies illustrated by level, competence and equipment

| Level 1 | Level 2 | Level 3 |
|---|---|---|
| Consumer level for self Authentication | Inspection level for Customs, banking, supply chain and retail inspection | Forensic for brand owner & laboratory use only |
| Also used for primary (first checks) by banking, customs and supply chain management | Checking authenticity if level one is ambiguous | Presenting evidence in court of law |
| Checking at this level requires only basic (but widespread) knowledge | Checking at this level requires a certain amount of confidential (covert) knowledge | Checking at this level requires a high degraee of skill and knowledge on a 'need to know' basis only |
| **Embossing and tactile inks such as intaglio**<br>**Watermarks in paper**<br>**Optically variable inks**<br>**Holograms and OVD's**<br>**Serialisation and encrypted codes**<br>**Watermarks in paper security threads**<br>**Colour change inks – heat rubbing etc** | **Encrypted RFID**<br>**Covert laser illumination for holograms and inks Quantum dots**<br>**Infra-red inks**<br>**Ultra-violet inks**<br>**Hidden image technology (HIT)**<br>**Photochromatic inks**<br>**Micro-printing and magnifyable taggants** | **Molecular markers**<br>**Nano-text**<br>**'DNA' type inks**<br>**Material biometrics** |
| **Verification tools**<br>Human senses and the use of SMS and smart phone apps to read bar codes or digital water-marks | **Verification tools**<br>UV lamps, IR Cameras, Decoding lens, magnifying glass or loupe, RFID tags with physically Un-clonable features<br><br>Access to closed proprietary databases | **Verification tools**<br>Laboratory tests including spectral analysis, microscopic observation, identification of nano-level features such as synthetic DNA or markers<br><br>Proprietary validation tools used for forensic validation—surface feature authentication |

**Figure 10.4 -** The importance of developing a holistic strategy that encompasses the various levels of security and the tooling necessary to authenticate each device

This process should be completed in conjunction with the perceived risks identified in the assessment process; counterfeiting, diversion, tampering etc.

## THE PRICE OF EVERYTHING AND THE VALUE OF NOTHING – COST VS. ROI

The playwright Oscar Wilde is famous for his observation that a cynic knows the price of everything and the value of nothing.

In the world of product security it should be recognized that no brand owner is likely to adopt a security system if it does not offer a return on investment (ROI). This fact of life can often be lost to those developing a new approach to protecting labels and packaging from product related crime.

Therefore, solutions that reduce or nullify the risks associated with tampering, counterfeiting, diversion et al will need to be measured against the savings or contribution they make to the overall well-being of the brand and also the consumer.

In a world where brands are constantly competing for market share and attempting to influence consumers through engagement and social media, it has never been more vital for brands to ensure that they keep their promises. Such promises may involve performance, safety and quality, all of which will be promoted on platforms such as Twitter and Facebook. Bad news can travel almost instantly through these channels, so brands need to be constantly aware of the risks they face if a crisis situation such as a counterfeit attack occurs.

Risk management will not only involve the assessment of consumer risk but also the effects that detrimental news such as a tampering or dilution event will have on overall value of the business. Shareholders are an important stakeholder in the value of a business and any event that reduces their return will also require a degree of consideration.

It should therefore be appreciated that it is not always possible to place an exact financial value on the extra investment necessary to protect a product from the risks identified. It is also possible that such risks themselves will only become evident after a successful initial attack and at that point the incentive to respond is much stronger and therefore easier to cost justify – especially if a claim for damages is taken up against the brand owner.

In order to gain a rough cost benefit analysis of a number of security features intended to address the three levels of security previously identified (overt, covert, forensic) the illustration (Figure 10.5) provides a useful indication of the level of security that can be achieved for a given cost. On the left axis the investment in security features climbs upwards from low through to high, and on the bottom axis the level of security provided by each alternative security technology increases from low to high the more you move to right. By combining two or more security features together the level of security obtained can be cost-effectively increased.

From this very basic model it should be possible to gain an idea of the relationship between various popular security devices and their respective cost and security level.

The savviest operators in this field will not be looking for the cheapest solution, but will be striving for the best solution.

For instance the best solution may well be a security label that combines the attributes of tamper evidence with authentication.

**Figure 10.6 -** Shows combined tamper evidence and authentication in a combined label

This could be extended further by introducing an

EAS circuit within the label to provide anti-theft protection too.

The incremental costs associated with each technology when applied to a single label, far outperforms the alternative which would require three separate labels. Not only would a saving be made in pressure-sensitive material but also a set off in manufacturing costs and the additional expense of affixing three labels to a pack rather than one.

Alternative solutions use labeling in conjunction with instruction on how to ensure that customers are purchasing authentic product in shops or online.

Murano Glass is a product that attracts knock offs, especially those purchased online, and since the process is unique to manufacturers in Murano, Venice and nowhere else. Pieces are individual and depending upon the design can cost upwards of hundreds of dollars.

By just searching for 'Murano Glass' on e-Bay it will be appreciated just how many fake products are being listed that claim this pedigree. Sadly the vast majority are cheap imports from China being sold by people who make more money out of postage and packing than they do from the glass products themselves.

By this point it will be apparent that anti-counterfeiting and other security devices that are

**Figure 10.8 -** The process of successful product security rests with informing all the stakeholders in the chain and educating each of them on the importance of correctly observing security features present on the product or via its packaging or labeling

**Figure 10.7 -** Trademark protection and authentication via a label can be backed up by an informative website

designed to provide tamper evidence, or as a protection or warning of other threats such as identifying illicit movement or theft, are not in themselves a comprehensive solution to the problem of suppressing product related crime.

Since packaging and labeling products are designed to 'Inform, Contain & Protect' (Chapter 1) goods they are the most useful and visible platform on which to set security monitoring devices. They are also ideal carriers of information that can convey supply chain detail such as track & trace data.

Such print related security can only function effectively as part of a carefully formulated strategy that involves systematically addressing every risk and providing the necessary information needed to every stakeholder in the process.

This will include retailers, distributors, customs/border protection, inspection teams and the final customer as well as the brand owner.

Since the Brand Owner is the primary driver of this process, it follows that the responsibility for introducing, controlling and informing all the stakeholders in the process lies with them.

Over the past quarter century, the solutions to securing and authenticating labels and packaging have evolved considerably and will need to advance continually in order to keep pace with the constant threats posed by the attractiveness of the financial incentives available to those who wish to exploit the opportunities offered through copying and tampering with branded goods.

Paradoxically, the greatest threat to branded products through counterfeiting and diversion lies in the widespread availability of fake goods on the internet. However, well-constructed and informative websites together with monitoring software are seen as useful tools with which to fight this criminality.

Similarly, inconsistency exists in the high number of solutions that are available to brand owners that recognise the need to tackle the problem. Some point to the fact that a form of standardization should exist, because a much smaller number of security devices would enable the development of recognizable 'standard' authentication systems available to everybody, rather than the requirement today that calls for an ever-widening range of verification tools and knowledge about how to use them.

Whilst this view is understandable, some would point out that the very strength of an effective authentication system lies in the difficulty in replicating it successfully and that standardization would offer fewer targets to compromise and thereby make life easier for those people determined to confront the system.

There is no simple answer to this problem, only that security resides in the exclusivity of the security devices chosen, whether this exclusivity resides in the process of manufacture of each device, or in the rareness or covertness of the materials involved in their fabrication.

What is certain though, is there will continue to be a requirement for well-designed and cost effective print related security systems and devices as long as the threats of product compromise continue to exist.

# Index

www.ingramcontent.com/pod-product-compliance
Lightning Source LLC
Chambersburg PA
CBHW041719210326
41598CB00007B/707